HOW TO . . .

Teach Mathematics
Using a Calculator

HOW TO . . .

Teach Mathematics Using a Calculator

Activities for
Elementary and Middle School

Terrence G. Coburn

National Council of Teachers of Mathematics

Copyright © 1987 by
THE NATIONAL COUNCIL OF TEACHERS OF MATHEMATICS, INC.
1906 Association Drive, Reston, Virginia 22091
All rights reserved

Third printing 1989

ISBN 0-87353-245-7

Printed in the United States of America

Table of Contents

Introduction

Recommendation 3 of the National Council of Teachers of Mathematics *Agenda for Action* states that "mathematics programs must take full advantage of the power of calculators and computers at all grade levels." This recommendation urges that students be provided access to calculators and that calculator use be integrated into the core mathematics curriculum on a routine basis, not as a replacement for all paper-and-pencil algorithmic skills, but as an enhancement to mathematics instruction.

This booklet provides examples of how the calculator can be used as an instructional tool in the elementary and middle school. It contains a brief discussion of some important instructional uses of the calculator, and considerations in selecting calculators for the classroom; activity pages, grouped by mathematical skill; teacher notes for each activity, including the purpose, appropriate grade range, calculator keys used, and an answer key; and sample tests with an answer key. The sample tests might be used for pre- and posttesting to see if students have learned to use the calculators and can recognize certain features of calculator use.

In addition to this booklet, the NCTM has published "Calculators," a focus issue of the *Arithmetic Teacher* (February 1987).

The Calculator as an Instructional Tool

The benefits of this powerful instructional tool will not be fully realized unless students have routine access to it. A calculator has speed, accuracy, decimal display, and memory, and it is a powerful computational device. These characteristics can be matched with objectives of the mathematics lesson. When any of the calculator's characteristics can serve in achieving the lesson's objective, then the calculator should be used. When the objective of the lesson is focused on paper-and-pencil skills, then the calculator can play a minor role or be placed out of view.

Some major uses of the calculator are described below:

1. To assist in the development of concepts

During group instruction the teacher can direct the use of the calculator to help interrelate the mathematical language, symbols, and representations used in developing a concept. The calculator is particularly helpful in presenting numeration concepts.

Example: Make the calculator count by tenths from 0 to 1. Coordinate the oral language used in counting, the shading of a rectangular unit diagram in tenths, and the pressing of the $\boxed{=}$ key.

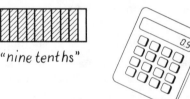

"nine tenths"

The student says "nine tenths," marks the rectangular unit region, and presses the $\boxed{=}$ key to see 0.9 on the display. Then the student says "ten tenths," marks and sees the whole unit, and thinks about the symbol he or she will see on the display when $\boxed{=}$ is pressed. In this manner, the student comes to understand that "ten tenths" and "one" are the same number.

2. To help reinforce skills

Many paper-and-pencil computation skills, estimation skills, and mental computation skills can be reinforced through the use of a calculator. Skill-reinforcement activities are usually conducted in individual or small-group situations following teacher-directed instruction.

Example: Students can read and translate word names for numbers into numerals. They can check their work with a check sum.

Read and enter correct digits:	*Press:*
Four hundred four	$\boxed{+}$
Four thousand forty	$\boxed{+}$
Forty-four hundred	$\boxed{=}$

Check number: 8844

3. To promote higher-level thinking

With a calculator a student can explore patterns and make conjectures. Readiness for generalizations can be fostered through working many examples. The calculator serves as a partner with the teacher, who poses the interesting questions and arranges the background and framework for the exploration.

$$37 \boxed{\times} 3 \boxed{=} \underline{\quad}$$
$$6 \boxed{=} \underline{\quad}$$
$$9 \boxed{=} \underline{\quad}$$
$$12 \boxed{=} \underline{\quad}$$
$$15 \boxed{=} \underline{\quad}$$
$$18 \boxed{=} \underline{\quad}$$
$$21 \boxed{=} \underline{\quad}$$

Example: Write a column of multiples of 3 from 12 to 99. Use a calculator to multiply each of these multiples by 37. Do you see a pattern to the products? Describe the pattern.

Students should be taught to make an organized exploration. The teacher should obtain a written description of a pattern from the students.

4. To enhance problem-solving instruction

The calculator can be used for the computation part of a problem, freeing the student to concentrate on the quantitative language in the problem.

Example: Select four numbers from the box on the right to fit in this problem. Use a calculator to check your selection.

Ruth purchased _____ adult tickets to the concert. The price of each ticket was _____ . She used a check made out for _____ dollars. How much money did she have left? Answer _____

$9.50
100
6
$43.00
$5.00

Types of Calculators

There are many makes and models of calculators available for less than five dollars each. Consider the following features when making your selection:

1. Most rudimentary four-function (+, −, ÷, ×) calculators come equipped with these keys:

   $\boxed{C}$ Clear: Clears the display, constant (if any), and arithmetic operation currently stored in the calculator.

   $\boxed{CE}$ Clear Entry: Clears the display only.

   $\boxed{CE/C}$ Some calculators combine the C and CE functions on one key. Press once and "Clear Entry" is performed. Press twice consecutively and "Clear" is performed.

   $\boxed{=}$ Equal: Business calculators combine "equal" with "plus" and "minus." Business calculators are not recommended for elementary school mathematics instruction.

2. Memory: Generally, four or five keys control the calcu-

lator's memory. The letter M usually appears on the display when a nonzero number is stored in memory.

$\boxed{M+}$ Adds the number currently on the display to the memory. The memory is cumulative.

$\boxed{M-}$ Subtracts the number on the display from the number in memory.

$\boxed{MC}$ Memory Clear: On most calculators, clears the memory but not the display.

$\boxed{MR}$ Memory Recall: Displays the total in memory. (Some calculators combine the MC and MR functions on one key.)

$\boxed{AC}$ All Clear: Some calculators also have this key, which clears the display, the memory, any constant, and any arithmetic operation.

3. Not all calculators have constants for addition and subtraction, features used in several of the activities in this booklet. The addition constant allows you to make the calculator "count" by repeatedly pressing the $\boxed{=}$ key. For example, pressing $2 \boxed{+} \boxed{=} \boxed{=} \boxed{=} \boxed{=} \ldots$ shows the multiples of two on the display.

   A subtraction constant allows students to count backward from a number: $10 \boxed{-} 1 \boxed{=} \boxed{=} \boxed{=} \boxed{=}$ shows 6.

4. Most calculators use multiplication and division constants. Generally, the first multiplication factor entered becomes a constant. For example, $3 \boxed{\times} \boxed{=} \boxed{=} \boxed{=} \boxed{=}$ shows 243 on the display, and $6 \boxed{\times} 4 \boxed{=} 7 \boxed{=}$ shows 24, then 42.

   The divisor becomes a constant: $24 \boxed{\div} 8 \boxed{=} 16 \boxed{=} 40 \boxed{=}$ shows 3, 2, and 5 in that order.

   Some brands of calculators have a $\boxed{K}$ key to invoke the constant feature. On other brands, the feature is activated by pressing the operation key twice consecutively.

5. $\boxed{+/-}$ Some calculators have a change-of-sign key. This key is useful when working with integers. When the key is pressed, the opposite of the displayed number is shown.

6. Solar power is generally more desirable than conventional battery power. However, there may be circumstances where available lighting conditions will not support solar power.

7. Calculator logic: Many nonscientific calculators perform arithmetic operations in the order in which they are entered. On these calculators, the display for the following key strokes would be 35: $3 \boxed{+} 4 \boxed{\times} 5 \boxed{=}$.

   However, there is a standard order of precedence in computing the value of such an expression. That is, multiplication and division are to be performed before addition and subtraction (unless parentheses are used). Most scientific calculators and some four-function calculators come equipped with algebraic logic. On these calculators the key strokes above would result in 23 as the answer.

Using the Activity Pages

The activities on the following pages illustrate some of the ways a calculator can be used at various grade levels. They are arranged by mathematical topic from simple counting activities to prealgebra and percent activities. The purpose and grade range for each activity are shown in the "Teacher Notes" at the end of this booklet. The teacher will have to study the purpose of the activity and look over the exercises to determine if the activity is suitable for the specific classroom in question. Some of the activities are two-person games (see "Guess My Number"). Others are meant for individual practice (see "Decimal Estimation") or small-group work (see "Shopping"). For some activities the teacher may teach the whole page to the class (see "Remainders" and "Memory").

A calculator for each student is desirable, though circumstances may dictate one calculator for every two students or some other ratio. It would be better to spread the calculator use throughout the school year than to bunch the experiences in one unit.

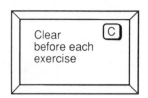

Clear
before each
exercise   Ⓒ

Counting

1. Count by ones: Press ⊞ 1 ⊟ ⊟ ⊟ . . . ⋅ ⋅ ⋅ *and so on*

   Stop at 50.   Time yourself.

   How long did it take to count to 50? _____

2. How long do you think it will take to count to 100?

   a. Write your prediction. _____

   b. Do it and time yourself. _____

3. How long do you think it will take to count from 1 to 200?

   a. Write your prediction. _____

   b. Do it and time yourself. _____

4. Count by twos. Press ⊞ 2 ⊟ ⊟ ⊟ . . .

   How long do you think it will take to count from 2 to 100?

   a. Write your prediction. _____

   b. Do it and time yourself. _____

5. Count by fives. Press ⊞ 5 ⊟ ⊟ ⊟ . . .

   How long do you think it will take to count from 5 to 100?

   a. Write your prediction. _____

   b. Do it and time yourself. _____

6. Count by tens. Press ⊞ 10 ⊟ ⊟ ⊟ . . .

   How long do you think it will take to count from 10 to 1000?

   a. Write your prediction. _____

   b. Do it and time yourself. _____

Counting On and Counting Back

1. Make your calculator count on from five.

Press 5 ⊞ 1 ⊟ ⊟ ⊟ ⊟ . . .

Stop when you reach 13.

How many times did you press ⊟? _____

Clear before each exercise ⓒ

2. Start at 25 and count by ones.

25 ⊞ 1 ⊟ ⊟ ⊟ . . .

Stop when you reach 35.

How many times did you press ⊟? _____

3. Start at 93. Close your eyes while you count by ones until you reach 100.

93 ⊞ 1 ⊟ ⊟ ⊟ . . .   Did you go past 100?

How many times did you press ⊟ to reach 100? _____

4. Start at 496. Count by ones and write each number until you reach 501.

496 ⊞ 1 ⊟ ⊟ ⊟ . . .

How many times did you press ⊟ to reach 501? _____

5. Make your calculator count back from 10 by ones.

10 ⊟ 1 ⊟ ⊟ ⊟ . . .

Stop at 0. How many times did you press ⊟? _____

6. Count back from 20 by ones. Stop at 13.

20 ⊟ 1 ⊟ ⊟ ⊟ . . .

How many times did you press ⊟? _____

7. Start at 45. Count back by ones to 35.

45 ⊟ 1 ⊟ ⊟ ⊟ . . .

Did you stop at 35? How many times did you press ⊟?

8. Try to count back from 30 by fives. Stop at 0.

30 ⊟ 5 ⊟ ⊟ ⊟ . . .

How many times did you press ⊟? _____

6

Mixed Order

Write the sum for each problem without using a calculator. Then use a calculator to check your answers.

1.   5 hundreds
         3 tens
  +   2 ones

2.   8 tens
         6 hundreds
  +   3 ones

3.   2 ones
         0 tens
  +   5 hundreds

4.   12 hundreds
         5 tens
  +   2 ones

5.   6 tens
         0 ones
  + 20 hundreds

6.   2 tens
     90 hundreds
  +   3 ones

7.   4 tens
         0 hundreds
  +   3 ones

8.   6 ones
         2 tens
  +   1 hundred

9.   1 hundred
         3 tens
  +   1 one

10.   8 ones
        0 tens
  +   2 hundreds

11.   3 tens
        6 ones
  +   3 hundreds

12.   1 hundred
        9 ones
  +   8 tens

13.   4 thousands
        3 tens
        4 hundreds
  +   3 ones

14.   6 ones
        7 tens
        2 thousands
  +   3 hundreds

15.   2 tens
        3 hundreds
        4 thousands
  +   5 ones

Number Word Names

Enter each number and then check the sum.

1. Four hundred twenty-three   ⊞

     Four hundred three   ⊞

     Four hundred   ⊟

       Check number: 1226

2. Eight hundred ninety   ⊞

     Six hundred nine   ⊞

     Two hundred ninety-nine   ⊟

       Check number: 1798

3. Three thousand, five hundred eighty   ⊞

     Two thousand, one hundred twenty-two   ⊞

     Seven thousand, seventy   ⊟

       Check number: 12 772

4. Nine hundred   ⊞

     Four thousand   ⊞

     Two thousand, sixty-four   ⊟

       Check number: 6964

5. Thirteen thousand, eight hundred   ⊞

     Twenty thousand, five hundred nine   ⊞

     Sixteen hundred   ⊟

       Check number: 35 909

6. Twelve hundred twenty   ⊞

     Ninety-five hundred three   ⊞

     Eight hundred four   ⊟

       Check number: 11 527

Calculator Counting Problems

1. Make your calculator count from 5 to 31 by twos.

5 ⊞ 2 ⊟ ⊟ ⊟ . . .

How many times did you press ⊟? _____

Check your answer on the calculator:

your answer ⊠ "counting-by" number ⊞ starting number ⊟

The calculator display should show the number given at the end of the check line.

Check: 13 ⊠ 2 ⊞ 5 ⊟ 31 (Display)

2. Count from 7 to 49 by threes.

7 ⊞ 3 ⊟ ⊟ ⊟ . . .

How many times did you press ⊟? _____ Write your answer in the check.

Check: _____ ⊠ 3 ⊞ 7 ⊟ 49

3. Count from 15 to 100 by fives.

15 ⊞ 5 ⊟ ⊟ ⊟ . . .

How many times did you press ⊟? _____ Write your answer in the check.

Check: _____ ⊠ 5 ⊞ 15 ⊟ 100

4. Count from 13 to 90 by sevens.

How many times did you press ⊟? _____ Check your answer.

Check: _____ ⊠ 7 ⊞ 13 ⊟ 90

5. Count from 3 to 75 by fours.

How many times did you press ⊟? _____ Check your answer.

Check: _____ ⊠ 4 ⊞ 3 ⊟ 75

6. Count from 7 to 88 by nines.

How many times did you press ⊟? _____ Check your answer.

Check: _____ ⊠ _____ ⊞ _____ ⊟ 88

9

Addition and Subtraction

Problems 1–4: Find the total weight.

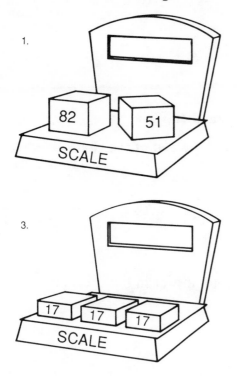

1. 82 51

2. 64 31 87

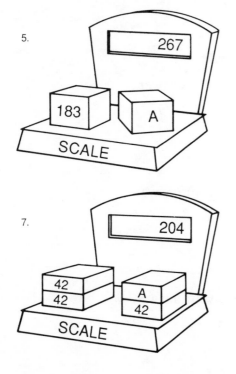

3. 17 17 17

4. 34 34 34 61 61 61 61

Problems 5–8: Find the weight of box A.

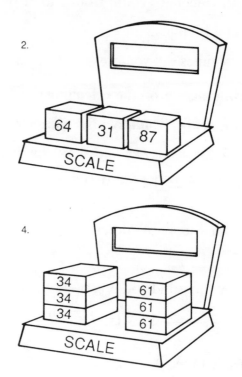

5. 267 — 183 A

6. 380 — 41 162 A

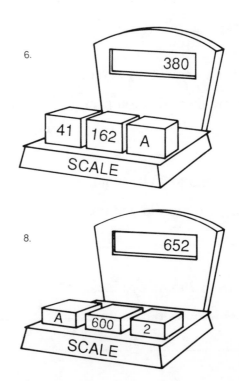

7. 204 — 42 42 A 42

8. 652 — A 600 2

10

Guess My Number

(Subtraction)

Two players
One calculator

One player starts the game by entering a secret number in the calculator. This is done by subtracting the number from itself. The display will show 0. The secret number should be written down (out of sight of the other player).

For example, press 13 ⊟ 13 ⊟ to hide 13 in the calculator.

Player 2 tries to find the number by using just the digit keys and the ⊟ key. The player should record each guess and the corresponding result. He or she will know that the secret number has been found when 0 shows on the display after the ⊟ key is pressed.

Game 1.	**Guess**	**Result**	*Game 2.*	**Guess**	**Result**
	_____	_____		_____	_____
	_____	_____		_____	_____
	_____	_____		_____	_____
	_____	_____		_____	_____

Game 3.	**Guess**	**Result**	*Game 4.*	**Guess**	**Result**
	_____	_____		_____	_____
	_____	_____		_____	_____
	_____	_____		_____	_____
	_____	_____		_____	_____

Note: The negative symbol shows on the display when the guessed number is smaller than the hidden number.

Estimating Sums

Circle each choice as quickly as you can. Then, after completing all the exercises, use the calculator to find the exact sums. Compare the exact sums with your choices.

Circle *one* choice

1. 196 + 184 + 209	Less than	500
	About	600
	More than	600
2. 95 + 142 + 233	Less than	400
	About	500
	More than	500
3. 352 + 367 + 348	Less than	1200
	About	1250
	More than	1300
4. 37 + 156 + 308	Less than	500
	About	550
	More than	600
5. 26 + 18 + 37 + 21	Less than	80
	About	100
	More than	120
6. 42 + 38 + 42 + 34	Less than	160
	About	200
	More than	240
7. 85 + 62 + 43 + 21	Less than	200
	About	210
	More than	220

Estimate the Quotient

Look at the division. Enter the divisor and press ⊠. Enter an estimate of the quotient and press ⊟. Revise your estimate twice. Subtract your last result from the actual dividend; the difference is your score. (Ignore the negative sign.) The lower the score, the better the estimate.

Sample:

62 ⊠ 120 ⊟

62)8792

130 ⊟

142 ⊟ (8804) ⊟ 8792 ⊟ 12
Score

Write each estimate.

1. 57 ⊠ _____ ⊟

57)9043

_____ ⊟

_____ ⊟ [ ] ⊟ 9043 ⊟

Score

2. 17 ⊠ _____ ⊟

17)2468

_____ ⊟

_____ ⊟ [ ] ⊟ 2468 ⊟

Score

3. 83 ⊠ _____ ⊟

83)604

_____ ⊟

_____ ⊟ [ ] ⊟ 604 ⊟

Score

4. 42 ⊠ _____ ⊟

42)74 681

_____ ⊟

_____ ⊟ [ ] ⊟ 74681 ⊟

Score

5. 13 ⊠ _____ ⊟

13)5007

_____ ⊟

_____ ⊟ [ ] ⊟ 5007 ⊟

Score

6. 36 ⊠ _____ ⊟

36)71 849

_____ ⊟

_____ ⊟ [ ] ⊟ 71849 ⊟

Score

7. 66 ⊠ _____ ⊟

66)1531

_____ ⊟

_____ ⊟ [ ] ⊟ 1531 ⊟

Score

8. 91 ⊠ _____ ⊟

91)403 625

_____ ⊟

_____ ⊟ [ ] ⊟ 403625 ⊟

Score

13

Division Estimation Game

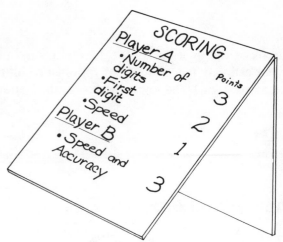

Three players
One calculator
List of division exercises. (Or use 3×5 cards with one exercise per card)

Player A is to write down a whole-number estimate for the quotient before Player B can display the actual answer on the calculator.
Player C uncovers the division exercise so that players A and B begin their tasks at the same time. Player C acts as referee and scorer. Rotate player positions to keep everyone involved.

Scoring: Player A receives 3 points for the correct number of digits in the whole-number estimate and 2 points for the correct first digit (the digit in the high-order place-value position). Player A also receives 1 point for writing an accurate estimate before Player B displays the actual answer on the calculator. Player B receives 3 points if he or she beats Player A.

Game 1

1. $67\overline{)48\ 307}$
2. 5 thousand divided by 41
3. $\dfrac{6789}{53}$
4. $86\ 900 \div 24$
5. $472\overline{)5016}$
6. 18 million divided by 2 thousand
7. $\dfrac{13\ 721}{643}$
8. $127\ 891 \div 513$
9. Divide 20 thousand by 875
10. $\dfrac{1}{18}$ of 6426

Game 2

1. 16 hundred divided by 12
2. $53\overline{)18\ 972}$
3. $\dfrac{1728}{88}$
4. $72\ 800 \div 36$
5. $521\overline{)4879}$
6. $309\ 421 \div 641$
7. $\dfrac{73\ 428}{855}$
8. $319\ 513 \div 641$
9. Divide 52 thousand by 921
10. $\dfrac{1}{24}$ of 5382

Guess My Number

(Division)

Two players
One calculator

One player starts the game by entering a secret number in the calculator. This is done by dividing the number by itself. The display will show 1. The secret number should be written down or stored in memory (out of the sight of the other player). For example, press 16 ÷ 16 = to hide 16 in the calculator.

The player who is trying to find the hidden number is restricted to using just the digit keys and the = key. The player should record each guess and the corresponding result. You may wish to speed the game by allowing the player to record the whole number and one or two decimal places of the result.

Sample Game (16 is hidden)

	Guess	Result
1.	50 =	3.1
2.	30 =	1.8
3.	10 =	.6
4.	15 =	.9
5.	16 =	1.

When the result is 1, the entered number is equal to the hidden number. You may wish to play with two-, three-, or even four-digit numbers.

Game 1:	Guess	Result	Game 2:	Guess	Result
1. _____		_____	1. _____		_____
2. _____		_____	2. _____		_____
3. _____		_____	3. _____		_____
4. _____		_____	4. _____		_____

Multiplication and Division

Patterns

Work three rows with a calculator. Write the answers. Write the answer in the fourth row without using a calculator. Check your answer with the calculator. Use the constant factor and constant divisor features of your calculator if possible.

1. 1001 ⊠ 82 ▣ _____
           65 ▣ _____
           14 ▣ _____
           57 ▣ _____

2. 2002 ⊠ 19 ▣ _____
           25 ▣ _____
           43 ▣ _____
           34 ▣ _____

3. 99 ⊠ 12 ▣ _____
           13 ▣ _____
           14 ▣ _____
           15 ▣ _____

4. 11 ⊠ 33 ▣ _____
           44 ▣ _____
           22 ▣ _____
           11 ▣ _____

5. 303 ⊠ 15 ▣ _____
           20 ▣ _____
           25 ▣ _____
           30 ▣ _____

6. 37 ⊠ 3 ▣ _____
           6 ▣ _____
           9 ▣ _____
           12 ▣ _____

7. 1 ⊠ 8 ⊞ 1 ▣ _____
      12 ⊠ 8 ⊞ 2 ▣ _____
    123 ⊠ 8 ⊞ 3 ▣ _____
  1234 ⊠ 8 ⊞ 4 ▣ _____

8. 101 ⊠ 22 ▣ _____
        222 ▣ _____
      2222 ▣ _____
    22222 ▣ _____

9. 37 ⊠ 36 ▣ _____
        39 ▣ _____
        42 ▣ _____
        45 ▣ _____

10. 9 ⊠ 9 ⊞ 7 ▣ _____
    98 ⊠ 9 ⊞ 6 ▣ _____
  987 ⊠ 9 ⊞ 5 ▣ _____
9876 ⊠ 9 ⊞ 4 ▣ _____

11. 11 ⊠ 11 ▣ _____
    111 ⊠ 111 ▣ _____
  1111 ⊠ 1111 ▣ _____
11111 ⊠ 11111 ▣ _____

12. 999999 ⊠ 2 ▣ _____
             3 ▣ _____
             4 ▣ _____
             5 ▣ _____

13. 4 ÷ 99 ▣ _____
  5   ▣ _____
  6   ▣ _____
  12   ▣ _____

14. 4 ÷ 999 ▣ _____
  7   ▣ _____
  3   ▣ _____
  12   ▣ _____

15. 2 ÷ 11 ▣ _____
  3   ▣ _____
  4   ▣ _____
  5   ▣ _____

16

Inverse

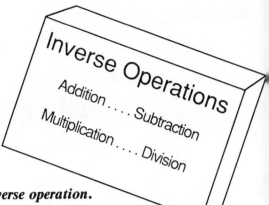

Check these problems. Some of the answers are wrong. Use the *inverse operation*.

1.
$$\begin{array}{r} 523 \\ 89\overline{)46\ 547} \end{array}$$

Press 89 ⊠ 523 ▣

2.
$$\begin{array}{r} 15\ 842 \\ \times\quad 297 \\ \hline 4\ 705\ 074 \end{array}$$

Press 4705074 ÷ 297 ▣

3.
$$\begin{array}{r} 83 \\ 365\overline{)32\ 485} \end{array}$$

4.
$$\begin{array}{r} 31\ 580 \\ \times\quad 592 \\ \hline 18\ 695\ 360 \end{array}$$

5.
$$\begin{array}{r} 612 \\ 909\overline{)556\ 308} \end{array}$$

6.
$$\begin{array}{r} 315\ 800 \\ \times\quad 592 \\ \hline 1\ 869\ 536 \end{array}$$

Find the hidden number. Start with the number on the right side of the equal symbol. Do the inverse operation.

7. [Hidden number] × 127 = <u>2159</u>

Press 2159 ÷ 127 ▣

Answer _____

8. [Hidden number] ÷ 679 = <u>529</u>

Press 529 ⊠ 679 ▣

Answer _____

9. [Hidden number] × 505 = <u>11 615</u>

Answer _____

10. [Hidden number] ÷ 713 = <u>70 587</u>

Answer _____

11. The product of some number and 59 is 5605. What is the unknown number?

_____

12. Some number divided by 168 has a quotient of 861. What is the unknown number?

_____

Counting by Tenths and Hundredths

1. Make your calculator count by tenths. Say the decimal name before you see the number on the display. Write the decimals as you count.

   Press ⊞ .1 ⊟ ⊟ ⊟ ⊟ ⊟   Write as you count aloud.

   .1 _____   _____   _____   _____   _____   [ ][ ][ ][ ][ ][ ][ ][ ][ ][ ]

                                                    tenths

   _____   _____   _____   _____   _____

   _____   _____   _____   _____   _____   Stop at 1.5

2. What is another name for ten tenths? Write the number that the calculator shows for ten tenths.

   _____

3. What is another name for twelve tenths? Write the number that the calculator shows for twelve tenths.

   _____

4. Write the word name for 14 tenths. _____ and _____ tenths.

5. Make your calculator count by hundredths. Say the decimal name before you see the number on the display.

   Press ⊞ .01 ⊟ ⊟ ⊟ ⊟   Count aloud. Stop when your display shows 0.09. What

   number will come next? Write your answer here _____ , then press ⊟ .

   hundredths

6. Continue counting by hundredths until your display shows 0.19. Write the number you think you'll see next.

   _____

   Continue counting. Stop when your display shows 0.99.

7. What number comes after ninety-nine hundredths when counting by hundredths? Write your answer here.

   _____

Notice that when you count by hundredths, you say twenty hundredths and the calculator shows 0.2 (two-tenths). Write what the calculator will show for each of these decimals.

8. Forty hundredths _____   9. One and twenty hundredths _____

10. Fifty hundredths _____   11. Two hundred hundredths _____

Writing Ratios as Decimals

Use a calculator to change the unit fraction to a decimal. Write the decimals for the fractions before you use a calculator to see the decimal. Use reasoning and mental computation.

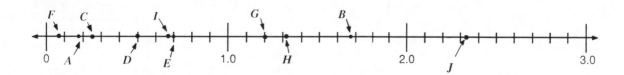

Press 2. Think 2 times 2 tenths. Write the decimal, then press ⊟.

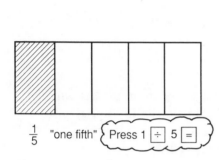

$\frac{1}{5}$ "one fifth" Press 1 ÷ 5 =

1. $\frac{2}{5}$ = _____ 2. $\frac{3}{5}$ = _____ 3. $\frac{4}{5}$ = _____

4. $\frac{5}{5}$ = _____ 5. $\frac{6}{5}$ = _____ 6. $\frac{7}{5}$ = _____

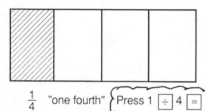

$\frac{1}{4}$ "one fourth" Press 1 ÷ 4 =

7. $\frac{2}{4}$ = _____ 8. $\frac{3}{4}$ = _____ 9. $\frac{4}{4}$ = _____

10. $\frac{5}{4}$ = _____ 11. $\frac{6}{4}$ = _____ 12. $\frac{7}{4}$ = _____

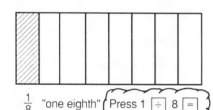

$\frac{1}{8}$ "one eighth" Press 1 ÷ 8 =

13. $\frac{2}{8}$ = _____ 14. $\frac{3}{8}$ = _____ 15. $\frac{4}{8}$ = _____

16. $\frac{5}{8}$ = _____ 17. $\frac{6}{8}$ = _____ 18. $\frac{7}{8}$ = _____

19. $\frac{8}{8}$ = _____ 20. $\frac{9}{8}$ = _____ 21. $\frac{10}{8}$ = _____

Look at the letters marking the locations of decimals on the number line. Write each letter next to the corresponding fraction. Check each answer with your calculator.

```
        F   C           I           G           B
        ↓  ↓            ↓           ↓           ↓
    +--•-•-•-•--+--•--•-+--•-+--•--+--•--+--•-+--•-+--•--+
    0    A    ↑    D  ↑        1.0    ↑         2.0    ↑        3.0
                                     H                J
                          E
```

22. $\frac{7}{10}$ _____   23. $\frac{1}{6}$ _____   24. $\frac{2}{3}$ _____   25. $\frac{4}{16}$ _____   26. $\frac{12}{10}$ _____

27. $\frac{3}{6}$ _____   28. $\frac{7}{3}$ _____   29. $\frac{5}{3}$ _____   30. $\frac{1}{16}$ _____   31. $\frac{8}{6}$ _____

Decimal Estimation

Enter the two factors into your calculator but do not press ⊟. Circle your choice for the best estimate of the product. Then press ⊟ to see if your choice was a good one.

1.	5.8	0.42	2.	13.2	0.65	3.	8.19	0.06
	× 7.2	4.2		× 5.1	6.5		× 0.673	0.6
		42.			65.			6.
		420.			650.			60.

In the following problems, enter the first factor into your calculator. Think of a whole number that will give a product that "lands" within the given range. Write this estimated factor down and then enter it into your calculator. Press ⊟ to see if your estimate is a good one. If you miss on your first estimate, enter a revised estimate and press ⊟. The limit is three tries. Write each factor you try before you press ⊟.

	Estimate	Range			Estimate	Range
4.	.8 × _____ =	125 . . . 150		**5.**	.33 × _____ =	200 . . . 225
	_____ =				_____ =	
	_____ =				_____ =	

	Estimate	Range			Estimate	Range
6.	4.27 × _____ =	80 . . . 92		**7.**	.48 × _____ =	425 . . . 435
	_____ =				_____ =	
	_____ =				_____ =	

	Estimate	Range			Estimate	Range
8.	1.96 × _____ =	357 . . . 365		**9.**	.134 × _____ =	800 . . . 815
	_____ =				_____ =	
	_____ =				_____ =	

	Estimate	Range			Estimate	Range
10.	23.6 × _____ =	1000 . . . 1500		**11.**	.07 × _____ =	50 . . . 55
	_____ =				_____ =	
	_____ =				_____ =	

Problems and Applications

Remainders

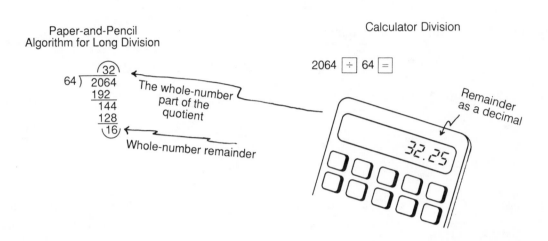

Paper-and-Pencil
Algorithm for Long Division

Calculator Division

2064 ÷ 64 =

The whole-number
part of the
quotient

Whole-number remainder

Remainder
as a decimal

Here is one approach for determining the whole-number remainder when you divide with a calculator:

(1) Press 2064 ÷ 64 =.
(2) Read and remember the whole-number part of the display, 32.
(3) Press 32 × 64 − 2064 =.
(4) The remainder is 16. Ignore the negative sign on the display. (Notice that 16 ÷ 64 = 0.25.)

Some of the following problems can be solved by finding the whole-number remainder. Use your calculator.

1. You have 876 marbles to pack in bags of 16 marbles each. How many marbles will be left over after you fill as many bags as you can?

2. Six hundred Boy Scouts will travel to camp on buses. Each bus will carry 64 scouts. How many buses will be needed to carry all of the scouts?

3. A machine places 35 cherries in each box. One million cherries will be packed. How many boxes will be needed?

4. You have to place 12 candy bars in each box. You must pack 127 boxes. How many candy bars will be left over if you have 1530 candy bars to pack?

5. Suppose that each time your bicycle wheels make one full revolution, you travel 6.3 feet. How many full revolutions do your wheels make while traveling one mile (5280 feet)?

6. How many full 7-day weeks are there in one million days and how many days are left over?

Average

One type of average for a set of numbers is the arithmetic mean. To find the mean of a set of numbers, you divide the sum of the numbers by the number of numbers.

Example: Find the mean of these weights:

$$6.4 \text{ kg} \qquad 8.2 \text{ kg} \qquad 10.9 \text{ kg} \qquad 6.8 \text{ kg}$$

Press 6.4 ⊞ 8.2 ⊞ 10.9 ⊞ 6.8 ÷ 4 ⊟ 8.075 (round to 8.1 kg)

Find the mean of each set of numbers.

1. Bowling Scores:

Debbie	110
Art	97
Frank	122
Alice	143

2. Test Scores:

68	43	98	81	52
46	81	77	54	67
33	92	85	57	56

3. Snow depth (in inches) on January 30 for eight years:

7	11.5	6	9.25
42	13.1	5	7.8

4. Heights (cm):

170.2	166.4
113	170
172.8	124.6
184	163
137	121.5
173	152.7

5. Weekly Earnings:

$21.80	$13.16	$31.25
$ 5.18	$ 7.20	$25.
$17.82	$11.07	$16.
$30.80	$24.15	

6. Ages of everyone at the family reunion:

68	92	11	5
43	47	51	16
8	14	22	63
9	2½	6 mos.	
39	19	89	

7. Points scored at each game during the school team's basketball season:

Home team	57	62	80	67	44	52	81	93	68	77	58
Opposition	43	61	84	60	53	51	78	81	57	84	52

What was the average score for the home team? The opposition? Find the average difference in score. (Subtract each pair of scores, regardless of who won.) Round each answer to the nearest whole number.

8. Thirty-four students each guessed the age of their teacher. The table shows the number of students and their guesses.

Students	5	13	14	2
	80	60	50	40

What was the average "guessed age"?

9. Suppose you know that the average of 13 numbers is 501. Twelve of these numbers are given below. What is the missing number?

222, 613, 527, 801, 712, 488, 907, 321, 602, 117, 417, 584

Memory

On many calculators, four keys operate a special memory:

M+ adds the number shown on the display to the memory.

M– subtracts the number shown in the display from the memory.

MR displays (recalls) the accumulated total in the memory.

MC clears the memory (but, on some calculators, not the display).

Note: Some calculators combine MC and MR on one key. Also, some calculators use an "all clear" AC key.

Try these experiments and see if you display the given result. Clear both the memory and the display before each exercise.

1. $16 \times 7 + 17 \times 6$ Result: 214
Press 16 [×] 7 [=] [M+] 17 [×] 6 [=] [M+] [RM].

2. $55 \times 5 - 44 \times 4$ Result: 99
Press 55 [×] 5 [=] [M+] 44 [×] 4 [=] [M–] [MR]

3. $17 \times 7 + 16 \times 6 + 15 \times 5 + 14 \times 4$
$+ 13 \times 3 + 12 \times 2 + 11$
Result: 420

4. $114 \div 19 + 104 \div 13$
Result: 14

5. $1000 - 25 \times 25 - 14 \times 14 - 3 \times 3$
Result: 170

6. $15 \times 15 + 15 - 16 \times 16 + 16$
Result: 0

7. $25 \times 25 \times 25 - 25 \times 25 - 25$
Result: 14 975

8. $\dfrac{17}{25} + \dfrac{40}{64}$ Result: 1.305
Press 17 [÷] 25 [=] [M+] 40 [÷] 64 [=] [M+] [MR]

9. $\dfrac{514}{125} - \dfrac{20}{32}$ Result: 3.487
(Use [M–] after 20 [÷] 32.)

10. $\dfrac{627 + 84 \times 17}{137}$
Result: 15

11. Suppose you have 16 pages of stamps with 24 stamps on each page and 37 pages with 19 stamps on each page. How many stamps do you have in all?

12. There are 7 months with 31 days, 4 months with 30 days, and one month with 28 days. How many days in all?

Problems and Applications

Shopping

Find the total bill for each set of items. Use the memory keys: M+ MR MC **Clear your calculator before each exercise.**

CHECKERS
2 ⁹⁷

Example: Suppose you bought . . .

3 toothbrushes
2 boxes Band-Aid bandages
What is the total bill?
Press   3 ⊠ .97 M+
        2 ⊠ 1.98 M+
        RM
Display shows 6.87

Band Aid
1 ⁹⁸

Toothbrush
97¢

+ AA
Batteries
1 ³⁵

1. Suppose you bought . . .

  2 packages batteries
  1 roll duct tape
  4 ChapSticks

99¢
Duct Tape

Chap Stick
2 for 99¢

2. Suppose you bought . . .

  5 boxes trash bags
  2 boxes cocoa mix
  3 boxes crayons

Total bill _____

Mello Cookies
5 for 1 ⁰⁰

Glue
1.25 fl. oz.
2/86¢

Total bill _____

3. Suppose you bought . . .

  1 ChapStick
  6 dozen clothespins
  2 bottles bath oil

Total bill _____

trash bags
10
1 ¹⁹

Clothes Pins
6 dozen 1 ⁴⁰

4. Suppose you bought . . .

  4 checker games
  4 bottles of glue
  4 Mello cookies

Total bill _____

Use M– **to calculate your change.**

5. Suppose you have $10.00. (Press 10 M+ .) You buy 7 boxes of trash bags. (Press 7 ⊠ 1.19 = M– .) What is your change? (Press RM .)

_____

sugar-free hot **cocoa mix**
8 envelope, 4.2 oz.
1 ⁷⁹

Crayons
1 ²⁷

Bath Oil 1 ¹³
32 oz.

6. Suppose you have $5.00. You buy 2 boxes of cocoa and 1 bottle of bath oil. What is your change?

_____

24

Ordering Fast Food

Six people eat lunch together each day of the week. Total the bill for each day.

Use the memory keys: M+ M− MR MC

1. Monday

    6 hamburgers
       (2 with cheese)
    3 large fries
    1 regular fries
    3 large colas
    1 milk   _____

    Use coupons whenever you can.

2. Tuesday

    3 fish sandwiches
    2 double hamburgers
    1 regular hamburger
       w/ cheese
    5 regular colas
    1 milk
    6 pies   _____

3. Wednesday

    13 regular hamburgers
     5 regular fries
     3 regular colas
     2 milks
     3 pies   _____

MENU

Hamburger		1.23
Double Hamburger		1.87
w/cheese		.15
Fish Sandwich		1.27
Fries	Large	.89
	Regular	.63
Cola	Large	.90
	Regular	.45
Milk		.47
Pie	Cherry	.72
	Apple	.72

Daily Coupon Specials

Wednesday	50¢ off hamburger
Thursday	1/2 off large fries
Friday	25¢ off large cola
Sat. & Sun.	1/2 off fish sandwich
Monday	Free regular fries
Tuesday	Free pie

4. Thursday

    5 double hamburgers
       (3 w/ cheese)
    1 fish sandwich
    8 large fries
    6 milks   _____

5. Friday

    2 regular hamburgers
       (1 w/ cheese)
    2 double hamburgers
       (1 w/ cheese)
    2 fish sandwiches
    3 large fries
    3 regular fries
    6 large colas   _____

6. How much money in all was spent on lunch for these 6 persons for the 5 days? _____

    What is the average amount spent by each person on each day? _____

Problems and Applications

Calculator Story Problems

Solve these story problems using your calculator. Use the (A), (B), and (C) numbers from the columns below each problem. Write your answer in the (D) column. Do each story problem three times, using a different set of numbers each time.

Problems 1–3:   Suppose you buy (A) tickets for the movies and each ticket costs you (B) . You had (C) . Now you have (D) left.

	A	B	C	D (Your answer)
1.	2	$3.75	$15.00	_____
2.	5	$2.25	$15.00	_____
3.	3	$4.50	$15.00	_____

Problems 4–6:   Suppose you start your savings with (A) . You plan to save (B) each week. It will take (C) weeks to save (D) .

	A	B	C	D (Your answer)
4.	$ 5.25	75¢	15	_____
5.	$ 3.15	65¢	25	_____
6.	$13.50	$1.50	19	_____

Problems 7–9:   Suppose your class is planning to buy a (A) . There are (B) students in your class. If each student brings (C) , there will be (D) left over.

	A	B	C	D (Your answer)
7.	$25.00 tree	24	$1.25	_____
8.	$58.00 globe	24	$2.85	_____
9.	$32.69 flag	35	95¢	_____

Problems 10–12:   Suppose you buy candy bars for (A) and sell them for a price of (B). Your profit on (C) candy bars is (D).

	A	B	C	D (Your answer)
10.	30¢ each	50¢ each	24	_____
11.	3 for $1.00	50¢ each	24	_____
12.	$3.30 a dozen	50¢ each	24	_____

26

Story Problems
(Fit the Facts)

Place the numbers from the box in the blank spaces in the problem. Use a calculator to check your decisions. Some exercises have more than one correct answer.

1. Fred read ____ books last year. He read ____ books this year. Fred says that he read ____ more books last year than this year.

| 41 | 218 | 31 | 259 |

2. Mollie has ____ toy cars. ____ are wooden and the rest are metal. She says that ____ cars are metal.

| 29 | 59 | 68 | 39 |

3. Karen dreamed that she bought and ate one chocolate ice-cream cone each day for a year. There are ____ days in a year. Each cone cost ____. She must have spent ____ in all.

| 365 | 15¢ | $292 | 80¢ |

4. Chad has saved ____ stamps. He keeps them in a book. There are ____ stamps on each page and there are ____ pages in the stamp book.

| 16 | 400 | 32 | 25 |

5. Hector's horse weighs ____ pounds. Nanci's horse weighs ____ pounds. If Hector's horse gains ____ pounds, it will weigh the same as Nanci's horse.

| 531 | 427 | 318 | 213 |

6. Ramone has lived for ____ days. His older brother has lived for ____ days. Ramone's older brother has lived for ____ days more than Ramone.

| 2190 | 3560 | 1825 | 4015 |

7. Sarah's family drove ____ days to California. They traveled ____ miles the first day. They traveled ____ miles each on the second and third days. The total distance driven to California was ____.

| 350 | 458 | 1158 | 3 |

8. Last week Thelma played in ____ basketball games. She scored ____ points in the first game. In the next game, she scored ____ more points than in the first game. Thelma scored ____ points in all.

| 15 | 91 | 2 | 4 | 38 |

27

Problems and Applications

Guess and Test
with a Calculator

1. Suppose you bought stamps to mail letters (22¢ each) and postcards (14¢ each). You spent a total of $7.74 on stamps. You mailed postcards and letters to 45 friends. How many stamps of each type did you purchase?

	22¢ stamps	14¢ stamps	Total cost	
Guess and test combinations that add to 45.	20	25	$7.90	Too high! Need to reduce the cost.

2. Suppose you sell 55 tickets to the play and collect a total of $176.00. The adult tickets sell for $4.50 each and the child tickets sell for $1.25 each. How many of each type of ticket did you sell?

	$4.50 ticket	$1.25 ticket	Total	
Guess some combination to make 55	25	30	150	Need more adult tickets!

3. Three people were discussing their birthdays. They discovered that the product of their ages (in years) was 3536. The product of the ages of the youngest (A) and oldest (C) was 221. The sum of the three ages was 46. What were their ages?

	Ages			Products	
	A	B	C	A × C	A × B × C
First guess	13	14	19		

4. A fancy ball-point pen (including its ink cartridge) costs $17.95. The pen costs $12.45 more than the ink cartridge. How much does the cartridge cost?

		Cartridge	Pen	Difference	
The cartridge and pen together cost $17.95	Guess	$5.00	$12.95	$7.95	Reduce the guess on the cost of the cartridge

5. A large pizza (16 slices) costs $8.60 and a small pizza (10 slices) costs $5.75. Suppose you spent $45.90 on pizza to feed 16 people. If each person ate 5 slices and there were 4 slices left over, how many pizzas of each size did you buy?

6. The perimeter of a certain rectangle is 656 feet. The area is 26 412 square feet. How much longer than wide is the rectangle?

7. A giant eight-pack of soda pop (including bottle deposit) costs you $7.84. The soda pop costs $5.12 more than the bottle deposit. How much is the deposit for each bottle?

28

Properties of Operations

1. If you change the <u>order</u> in adding two numbers, the sum remains the same. Try this example. Write each sum.

$$684 + 97 = \underline{\hspace{2cm}} \qquad\qquad 97 + 684 = \underline{\hspace{2cm}}$$

2. What about subtraction? Does changing the <u>order</u> of subtracting change the result? Yes or no?

$$508 - 492 = \underline{\hspace{2cm}} \qquad\qquad 492 - 508 = \underline{\hspace{2cm}}$$

What about multiplication or division? Try these examples and circle your answer.

3. Division: Order <u>does</u> matter.
      Order <u>does not</u> matter.

$$132 \div 16 = \underline{\hspace{2cm}}$$

$$16 \div 132 = \underline{\hspace{2cm}}$$

4. Multiplication: Order <u>does</u> matter.
       Order <u>does not</u> matter.

$$18 \times 55 = \underline{\hspace{2cm}}$$

$$55 \times 18 = \underline{\hspace{2cm}}$$

If there are three or more numbers in the expression, you must choose which two to compute first. Does changing your choice change the results? Explore with these expressions:

5. $84 + 91 + 67$

($84 + 91$ first) Final sum $\underline{\hspace{2cm}}$

($91 + 67$ first) Final sum $\underline{\hspace{2cm}}$

6. $271 - 58 - 13$

($271 - 58$ first) Final answer $\underline{\hspace{2cm}}$

($58 - 13$ first) Final answer $\underline{\hspace{2cm}}$

7. $19 \times 20 \times 21$

(19×20 first) Final product $\underline{\hspace{2cm}}$

(20×21 first) Final product $\underline{\hspace{2cm}}$

8. $1152 \div 96 \div 6$

($1152 \div 96$ first) Final quotient $\underline{\hspace{2cm}}$

($96 \div 6$ first) Final quotient $\underline{\hspace{2cm}}$

9. Write your thoughts on changing the grouping of numbers with each operation.

$\underline{\hspace{16cm}}$

$\underline{\hspace{16cm}}$

The distributive property of multiplication over addition states that the product of a number and a sum of two numbers is the same as the sum of the two products: $a(b + c) = ab + ac$. This property is illustrated in the following example:

$$16 \times (53 + 47) = \underline{\quad 1600 \quad}$$
$$16 \times 53 + 16 \times 47 = \underline{\quad 1600 \quad}$$

Match the expression on the left with the one on the right having the same value. You may use your calculator to decide.

10. $83 \times 51 + 51 \times 46$

11. $(51 + 83) \times 46$

12. $83 \times (51 + 46)$

a. $46 \times 51 + 46 \times 83$

b. $83 \times 51 + 46$

c. $51 \times (83 + 46)$

d. $83 \times 51 + 83 \times 46$

Use Reasoning to Find the Answer

You may use the calculator to work any three exercises in a game. You receive 10 points at the beginning of the game. You lose 5 points for each time you use the calculator beyond the three free uses. You <u>cannot</u> use paper and pencil to compute. Use your pencil to write answers only. You lose 2 points for each incorrect answer.

In each game, first study each of the six exercises. Then choose which three you will do with the calculator.

Game 1	Answer
1. 328 times 684 is	_____
2. The product of 849 and zero is	_____
3. Multiply 815 and 9774.	_____
4. $684 \times 328 = ?$	_____
5. $9774 \times 815 = ?$	_____
6. One times 3 million is	_____

Game 2	Answer
1. 357×842	_____
2. 42×357	_____
3. 800×357	_____
4. 358×842	_____
5. 357×800	_____
6. 357×42	_____

Game 3 $M = 8473, N = 972$	Answer
1. M plus N	_____
2. M times N	_____
3. N plus M	_____
4. Zero plus N	_____
5. $(900 + 72)$ times M	_____
6. $(8400 + 73)$ times $(900 + 72)$	_____

Game 4 $A = 575, B = 75, C = 500$	Answer
1. 6675 times A	_____
2. $(B + C)$ times 6675	_____
3. A minus C	_____
4. 6675 divided by B	_____
5. $6675 \times B + 6675 \times C$	_____
6. $(599 \times 681) - (681 \times 599)$	_____

Game 5 $A = 899, B = 901$	Answer
1. $A + B$	_____
2. $A \times B$	_____
3. $B - A$	_____
4. $B \times A$	_____
5. 100 times A	_____
6. $A \times 900 + A$	_____

Game 6 $M = 473, N = 2838$	Answer
1. N divided by M	_____
2. M times N	_____
3. Zero divided by M	_____
4. One times M times N	_____
5. Ten times M	_____
6. One hundred times N	_____

My Dear Aunt Sally

The phrase **My Dear Aunt Sally** serves as a reminder of the rule that **M**ultiplication and **D**ivision are to be performed before **A**ddition and **S**ubtraction when computations are done from left to right in an arithmetic expression. However, your calculator may be one that performs operations in the order they are entered.

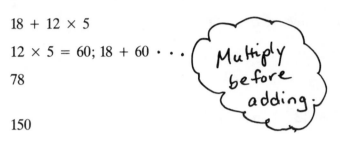

Arithmetic expression:	$18 + 12 \times 5$
Order of calculating:	$12 \times 5 = 60; 18 + 60 \cdots$
Correct result:	78
Result (incorrect) on some calculators:	150

Work these exercises with your calculator to see if you obtain the given results.

1. $15 \times 12 + 84 \times 12$
Result: 1188

2. $64 - 18 + 53 \times 11$
Result: 629

3. $208 \div 16 + 425 \div 17$
Result: 38

4. $85 \times 84 - 207 \div 23$
Result: 7131

Parentheses are used to indicate the order of computation. Perform operations within parentheses first.

5. $(293 + 213) \div (64 - 18)$
Result: 11

6. $(17 \div 136) + (28 \div 32)$
Result: 1

7. $4625 + 5264 \div (2188 - 872)$
Result: 4629

8. $871 - (52 \times 16) + 372$
Result: 411

9. $121 + (84 - 27) + 61 - (41 + 13)$
Result: 185

10. $(86 - 61) \times (86 + 61)$
Result: 3675

For each of the following, insert the operation symbols to give the results shown. Draw parentheses, if necessary, to show the order of operations.

+	−	×	÷

11. 53 ___ 18 ___ 10 = 964

12. 521 ___ 16 ___ 16 = 265

13. 127 ___ 92 ___ 112 ___ 83 = 24

14. 1325 ___ 184 ___ 131 = 25

31

Linear Equations

Solve this equation by working backward: $13N + 82 = 810$

Enter 810

Press $\boxed{-}$ 82 $\boxed{=}$ (the inverse of adding 82)

Press $\boxed{\div}$ 13 $\boxed{=}$ (the inverse of multiplying by 13)

The display shows 56.

Check the solution: 13 $\boxed{\times}$ 56 $\boxed{+}$ 82 $\boxed{=}$

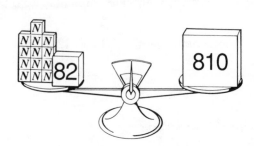

Solve. Check each solution.

1. $27N - 161 = 730$

2. $51N - 810 = 1587$

3. $85N + 65 = 1935$

4. $117N + 2244 = 20\,847$

5. $36N - 214 = 2954$

6. $13N - 987 = 1080$

7. $204N + 1489 = 2713$

8. $486N - 17\,833 = 7439$

9. $147 + 54N = 7923$

10. $7526 + 39N = 293\,513$

11. $3000 = 789N - 945$

12. $38\,512 = 456N - 350\,000$

13. $657N = 531\,513$

14. $N - 892 = 583$

15. $16N + 256 = 512$

16. $45N - 1234 = 476$

17. $753N - 18\,825 = 0$

18. $0 = 125N - 15\,625$

19. $\dfrac{N}{23} + 89 = 134$

20. $\dfrac{N}{47} + 547 = 630$

(The inverse of division is multiplication.)

Proportions

Solve this equation by multiplying first. Then divide.

$$\frac{N}{85} = \frac{27}{51}$$

$$N = \frac{27 \times 85}{51}$$

Enter 27

Press $\boxed{\times}$ 85

Press $\boxed{\div}$ 51 $\boxed{=}$

The display shows 45.

Use cross products to check the solution: 85 $\boxed{\times}$ 27 $\boxed{=}$ 45 $\boxed{\times}$ 51 $\boxed{=}$

Note that you can divide first, then multiply.

Enter 27   Press $\boxed{\div}$ 51   Press $\boxed{\times}$ 85

The display shows 44.999994. This can be rounded to 45. When you divide first, the displayed results may have to be rounded to the whole-number solution, so it is better to multiply first.

Solve. Check each solution.

1. $\frac{N}{27} = \frac{40}{72}$

2. $\frac{N}{245} = \frac{64}{98}$

3. $\frac{N}{512} = \frac{936}{768}$

4. $\frac{N}{256} = \frac{51}{272}$

5. $\frac{N}{48} = \frac{272}{51}$

6. $\frac{N}{272} = \frac{48}{256}$

7. $\frac{N}{182} = \frac{143}{169}$

8. $\frac{N}{143} = \frac{182}{154}$

9. $\frac{N}{154} = \frac{169}{143}$

Perform the operations in parentheses first. Round answers to the nearest whole number.

10. $\frac{N}{64} = \frac{(389 + 427)}{55}$

11. $\frac{N}{52} = \frac{(1206 - 389)}{427}$

12. $\frac{(139 + 625)}{150} = \frac{N}{17}$

13. $\frac{184}{(627 - 509)} = \frac{N}{38}$

33

Exponents

An *exponent* tells you how many times to use the *base* number as a factor.

A calculator can compute with exponents by using the constant-factor feature.

Press: 6 ☒ ▭ ▭

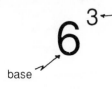

base

exponent

6^3 means $6 \times 6 \times 6$

$6^3 = 216$

Read "Six to the third power is 216."

Each time the equal key is pressed, the number on the display is multiplied by the constant factor.

Compute the value of each of these exponential expressions.

1. 7^3 _____ **2.** 4^5 _____ **3.** 2^8 _____ **4.** 10^3 _____

5. 11^3 _____ **6.** 6^5 _____ **7.** 2^{10} _____ **8.** 3^{10} _____

9. 4^{10} _____ **10.** 50^3 _____ **11.** $(16 \times 3)^2$ _____ **12.** $(32 + 57)^3$ _____

Try working exponents with decimal-number bases.

13. $(1.5)^3$ _____ **14.** $(2.4)^4$ _____ **15.** $(0.9)^4$ _____ **16.** $(0.5)^5$ _____ **17.** $(1.9)^{10}$ _____

18. Notice that when the base is less than 1, the value of an expression decreases as the exponent increases. Why do you think this is true?

Explore the pattern of the ones digit in each of these.

19. $2^1 =$ _____   **20.** $3^1 =$ _____   **21.** $4^1 =$ _____   **22.** $7^1 =$ _____

$2^2 =$ _____ $3^2 =$ _____ $4^2 =$ _____ $7^2 =$ _____

$2^3 =$ _____ $3^3 =$ _____ $4^3 =$ _____ $7^3 =$ _____

$2^4 =$ _____ $3^4 =$ _____ $4^4 =$ _____ $7^4 =$ _____

$2^5 =$ _____ $3^5 =$ _____ $4^5 =$ _____ $7^5 =$ _____

$2^6 =$ _____ $3^6 =$ _____ $4^6 =$ _____ $7^6 =$ _____
$\vdots$ $\vdots$ $\vdots$ $\vdots$

Solve for the exponent in each of the following:

23. $2^E = 32\ 768$ **24.** $7^E = 5\ 764\ 801$ **25.** $13^E = 4\ 826\ 809$

$E =$ _____ $E =$ _____ $E =$ _____

Integers

+/− The change-of-sign key is helpful in computing with positive and negative numbers.

Example: $^+16 + ^-9$

Press 16 ⊞ 9 +/− ⊟   *7*

A negative number is entered by entering the number and then pressing the change-of-sign key. Notice the negative symbol in the display. A positive number does not show a positive symbol.

1. Add three negative numbers.

$^-27$ $^-18$ $^-51$

Sum _____

2. Add 53 to a negative 18.

$53 + ^-18$

Sum _____

3. Subtract 61 from 43.

$43 - 61$

Difference _____

Now add $^-61$ to 43.
Is the result the same?

4. Subtract $^-18$ from 42.

$^+42 - ^-18$

Difference _____

Now add 18 to 42.
Is the result the same?

5. Add 45 and $^-45$.

Sum _____

(45 and $^-45$ are called opposites)

6. What number is added to $^+69$ to make a sum of zero?

Answer _____

7. a. $17 - ^-53 = $ _____

b. $17 + 53 = $ _____

8. a. $^-24 - 16 = $ _____

b. $^-24 + ^-16 = $ _____

9. a. $41 - 57 = $ _____

b. $41 + ^-57 = $ _____

10. Study exercises 7–9 and complete this statement: When subtracting some number B from some number A, you can _____ the opposite of B to A.

11. Multiply $^-8$ and $^-7$

Product _____

12. Divide $^-408$ by 17.

Quotient _____

13. Do you think that the product of two negative numbers will always have a positive product? Check your conjecture with a few more examples.

Evaluate each of these expressions:

14. $(^-64 + 85) \times ^-12$ _____

15. $^-25 \times (19 - 37)$ _____

Perimeter

The *perimeter* of a polygon is the total of the lengths of all its sides.

Find the perimeter of each polygon.

1.

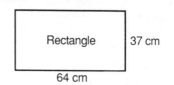

2.

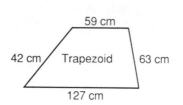

3.

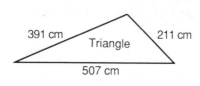

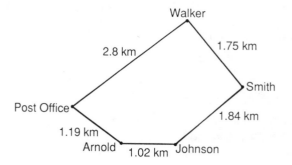

4. The letter carrier's mail route is in the shape of a pentagon. How far does he travel on his route? Start at and return to the post office.

5. The 8 sides of a *regular* octagon are congruent. If the length of one side is 632 cm, what is the perimeter?

6. The 5 sides of a *regular* pentagon are congruent. If the length of one side is 0.86 km, what is the perimeter?

7. The perimeter of a *regular* triangle (equilateral) is 135 cm. How long is each of the sides?

8. Two of the three sides of a triangle have a sum of 958 cm. What is the length of the third side if the perimeter is 1207 cm?

9. The short side of a rectangle is 19 cm long. The length is twice the width. What is the perimeter?

10. A strip of cloth 8 cm wide is to be sewn along the perimeter of a rectangular tablecloth. If the tablecloth is 533 cm long by 243 cm wide, how long a strip is needed?

11. A certain baseball player hit 660 home runs during his career. Suppose he ran about 90 feet from base to base. About how many miles (to the nearest tenth) did he run while hitting home runs (5280 feet equals one mile)?

Area

The calculator formulas for computing the areas of certain shapes are given in the box below.

Find the area of the following polygons:

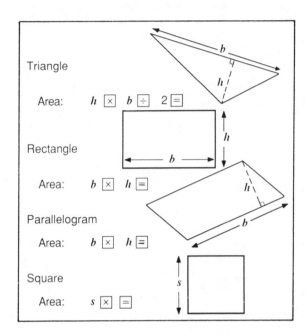

1. Triangle
   Height = 35 cm
   Base = 72 cm

   _____

2. Rectangle
   Base = 184 cm
   Height = 95 cm

   _____

3. Parallelogram
   Base = 203 cm
   Height = 59 cm

   _____

4. Triangle
   Base = 6.25 m
   Height = 54 m

   _____

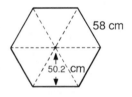

5. This regular hexagon is made up of 6 triangles. Each base is 58 cm and each height is 50.2 cm. What is the total area of the hexagon?

   _____

6. A certain triangle has a base of 84 cm. Its height is twice the base. What is its area? _____

7. A certain rectangle is 21 cm by 28 cm. The dimensions of another rectangle are three times as large. How much more area does the larger rectangle have than the smaller rectangle?

   _____

8. A rectangular prism has 6 rectangular faces. The 2 end faces are squares, each with sides of 62 cm. The other 4 faces are rectangles, 62 cm by 115 cm. What is the total surface area of this rectangular prism?

   _____

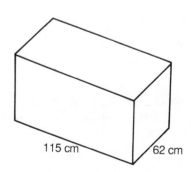

37

Circles, Spheres, Cones, and Cylinders

The famous constant named with the Greek letter π (pi) is the ratio of the circumference of a circle to its diameter. Try this experiment:

Step 1. Take a sheet of 8.5-by-11-inch paper and roll it to form a cylinder. Tape it. The circumference of the circular top (and bottom) is 11 inches.

Step 2. Measure the diameter of the circular end(s).

Step 3. Divide the circumference 11 by the estimated diameter. The ratio should be close to 3. Try this with different-sized pieces of paper.

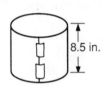

The approximate value of π is 3.14159265. Below are some calculator formulas that use π.

Circumference of a circle:

$\pi \boxtimes d \boxminus$

Area of a circle:

$r \boxtimes \boxminus \boxtimes \pi \boxminus$

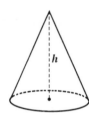

Volume of a right circular cone:

$r \boxtimes \boxminus \boxtimes \pi \boxtimes h \boxdiv 3 \boxminus$

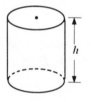

Volume of a right circular cylinder:

$r \boxtimes \boxminus \boxtimes \pi \boxtimes h \boxminus$

Volume of a sphere:

$r \boxtimes \boxminus \boxminus \boxtimes \pi \boxtimes 4 \boxdiv 3 \boxminus$

Use 3.14 for π in the following computations:

1. Find the area of a circle with a radius of 15 cm.

2. Find the volume of a right circular cylinder with a radius of 25 cm and a height of 36 cm.

3. Find the volume of a right circular cone having the same radius and height as the cylinder in exercise 2.

4. The inside of a can of 3 tennis balls is about 23 cm high and has a diameter of about 7.5 cm. Each ball has an outer circumference of about 23 cm. How much space is left in the can?

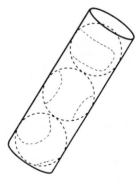

Change a Ratio to a Percent

| % | Chad got 16 questions correct out of the 25 questions on the test. His teacher marked 64% at the top of the test paper. Check this score using your calculator's percent key:

Press 16 $\div$ 25 %

The % key is used in place of $=$. It positions the decimal point to show the number of hundredths. Percent means hundredths.

Write the percent for each problem. Round answers to the nearest whole-number percent.

1. Ellen made 9 out of 12 free throws in the basketball game. She shot _____ percent.

2. Hector saved $5.94 out of his $39.60 paycheck. He saved _____ percent of his pay.

3. Glenda was in school on 162 days out of the total 180 school days. She attended _____ percent of the school days.

4. 9 out of every 15 students voted to go to the ballet. What percent voted to go to the ballet?

5. The average adult in Jonesville spends 9 of every 24 hours watching television. What percent of 24 hours is this?

6. There are 21 boys and 17 girls in the class. What percent of the total class are girls? What percent of the total class are boys?

Note: Some of the following results are less than 1 percent or greater than 100 percent. Round answers to the nearest tenth of a percent when necessary.

7. Compare Gary's weight now at age 18—203 pounds—to his weight of 98 pounds when he was 12 years old. What percent of his 12-years-old weight is his present weight?

8. Chad now earns $12 each time he cuts the lawn. Two years ago he earned $7 for mowing the lawn. What percent of his old pay is his present pay?

9. Suppose you spent the following amounts of money on a Saturday:

2 movie tickets	$4.75 each
2 popcorn	2.25 each
2 candy bars	1.15 each
2 sodas	1.75 each
1 phone call	.20
1 T-shirt	9.99

The cost of the phone call was _____ percent of the total amount of money you spent on Saturday.

10. Mr. Smith put 118 527 miles on his old car before he sold it. Part of those miles was a 258-mile trip to Chicago. What percent of the total mileage was the Chicago trip?

Several Percents of the Same Base

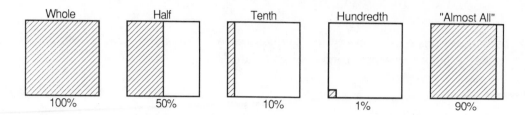

Whole	Half	Tenth	Hundredth	"Almost All"
100%	50%	10%	1%	90%

Calculate 50 percent, 10 percent, 1 percent, and 90 percent of the same base amount. Write the percentages as you press %.

Enter the base amount (the whole) first. Use your calculator's constant-factor feature (if available) to speed up your calculations.

	Base	Enter	Write		Base	Enter	Write
1.	$300.00	300 ⊠ 50 %	_____	**2.**	$80.00	80 ⊠ 50 %	_____
		10 %	_____			10 %	_____
		1 %	_____			1 %	_____
		90 %	_____			90 %	_____
3.	$650.00	650 ⊠ 50 %	_____	**4.**	$75.00	75 ⊠ 50 %	_____
		10 %	_____			10 %	_____
		1 %	_____			1 %	_____
		90 %	_____			90 %	_____
5.	$5.00	5 ⊠ 50 %	_____	**6.**	$426.00	426 ⊠ 50 %	_____
		10 %	_____			10 %	_____
		1 %	_____			1 %	_____
		90 %	_____			90 %	_____
		100 %	_____			100 %	_____
		*150 %	_____			*150 %	_____
7.	$18.60	18.6 ⊠ 50 %	_____	**8.**	$0.84	.84 ⊠ 50 %	_____
		10 %	_____			10 %	_____
		1 %	_____			1 %	_____
		90 %	_____			90 %	_____
		100 %	_____			100 %	_____
		*200 %	_____			*200 %	_____

*More than the whole. For example, 150% means 100% and 50% more.

Estimating a Percentage Using the Same Rate

100% of $6.50 is the whole $6.50. Estimate 90% of $6.50. Use the constant-factor feature (if available) to speed up your calculations.

Step 1. Press 90 ⊠ 6.50
Step 2. Think: 90% is more than half of $6.50—almost all of it.
Step 3. Circle your choice:

$0.58 $5.85 $58.50

Step 4. Press ▣

Estimate 90% of each of the amounts in problems 1–3. Use your estimate to help choose the correct answer from the amounts given below each problem. Then use your calculator to see if your choice was correct.

1. 90% of $75.00

$ 0.68
$ 6.75
$67.50

2. 90% of $39.64

$356.76
$ 35.68
$ 3.57

3. 90% of $125.90

$ 0.11
$113.31
$ 11.33

Estimate 48% of each of the amounts in problems 4–6. Choose the correct answer, then check your choice with the calculator.

4. 48% of $900.00

$ 0.43
$ 4.32
$ 43.20
$432.00

5. 48% of $15.00

$ 72.00
$ 7.20
$ 0.72
$720.00

6. 48% of $63.75

$ 30.60
$ 306.00
$ 3.06
$3060.00

Estimate 9% of each of the amounts in problems 7–12. Choose the correct answer, then check your choice with the calculator.

7. 9% of $340.00

$ 0.31
$ 30.60
$ 3.06
$306.00

8. 9% of $1407.00

$126.63
$ 12.66
$ 1.27
$ 0.13

9. 9% of $24.62

$221.58
$ 22.16
$ 0.22
$ 2.22

10. 9% of $8100.00

$ 7.29
$ 72.90
$ 729.00
$7290.00

11. 9% of $6.15

$ 5.54
$ 0.55
$ 0.06
$55.40

12. 9% of $62.50

$ 5.63
$ 0.56
$56.25
$ 0.06

Twelve Is Fifty Percent of What?

Marion is 12 years old. Her mother told her that her age was 50 percent of her brother's age. Marion thought: 12 is 50% of how much? This means the ratio of my age to my brother's age is 50%. I'll divide to find my brother's age.

12

┌─────┬─────┐
│░░░░░│   │
│50%░░│   │
│░░░░░│   │
└─────┴─────┘

Brother's age
is the whole
(base)

$12 \boxed{\div} 50 \boxed{\%}$ _____

Check: $50 \boxed{\times} 24 \boxed{\%}$
50% of 24 is 12

Find the whole amount (base) in each problem. Multiply to check.

1. Brian's age is 20% of his father's age. If Brian is 7 years old, how old is his father?

2. Sarah has saved 80% of the money she needs to buy a new jacket. If she has saved $24.00, how much does the jacket cost?

3. Philip got 90% correct on his test. If he answered 27 questions correctly, how many questions were on the test?

4. State Senator Jackson received 413 724 votes. If this number is 33 percent of the total vote, how many votes were cast in all?

5. Steve made 60% of his shots in the basketball game. If he made 3 baskets, how many shots did he attempt?

6. Fifteen percent of the students are absent from school today. If 595 students are in school today, how many students are absent?

Be careful here.

Find the unknown amount. Round to the nearest cent if necessary. Multiply to check.

7. 12.5% of how much is $52.00?

8. 16.75% of how much is $40.00?

9. 87.5% of how much is $36.50?

10. 33.3% of how much is $9.15?

11. 0.97% of how much is $2.25?

12. 49.9% of how much is $0.36?

Add-ons and Markups

Suppose the grocery bill was $46.20 plus 5% tax.

Press 46.2 ⊞ 5 %. The total bill is $48.51.

Note: On some calculators you will see 2.31 when you press 5 %; you must then press
⊟ to obtain the sum of 46.20 and 2.31.

A 40% markup on the price of this watch can be calculated two different ways.

40% Markup	
Store Cost	Customer Cost
$8.45	_____

Method 1. Think of adding 40% to the original price.

8.45 ⊞ 40 %

Total: $11.83

Method 2. Think: 100% plus 40% is 140%.

140 ⊠ 8.45 %

Total: $11.83

Calculate each amount.

1. $65 ⊞ 80 %

2. $120 ⊞ 5 %

3. $92 ⊞ 4.5 %

4. 180% of $65

5. 105% of $120

6. 104.5% of $92

Solve these problems.

7. Joan wants to leave a 15% tip. Her lunch bill was $8.75. How much will the lunch and tip cost altogether?

8. Next year John wants to deliver 80% more newspapers each day than he delivers this year. If he delivers 210 papers now, how many does he plan to deliver each day next year?

9. The hotel charges 7.5 percent recreation tax. How much will a $54 room cost with this added tax?

10. The meal costs $16.24. Add to this a 6% tax and a 20% tip. What is the total cost of the meal?

Student Evaluation

The sample tests on the following pages are related to the objectives listed below. Test 1 is suitable for primary grades, Test 2 for upper elementary, and Test 3 for middle school. Part of each of Tests 2 and 3 should be taken without a calculator in view. The calculator features assumed on some items include an automatic constant for addition and multiplication.

These tests are meant to be suggestive. They have not been validated, and norms have not been established.

Objectives

1. To use the calculator to perform rudimentary computations

   Examples

   A. Multiply 807 and 59   B. $\dfrac{34.2 \times 68}{15}$

2. To recognize specific keys and apply them correctly

   Example (calculator not in view)

   Which number will be shown on the display after these keys are pressed?

   $\boxed{C}\ 6\ \boxed{+}\ 8\ \boxed{CE}\ 3\ \boxed{=}$

   a. 9   b. 11   c. 14   d. 17

3. To read and interpret the calculator display

   Example (calculator allowed)

   Suppose you multiply $2.25 by 42 on the calculator. Which of the following is the answer in dollars and cents?

   a. $9.45   b. $94.50   c. $945.00   d. $9450.00

4. To recognize and apply specific features of the calculator such as automatic constants

   Example (calculator not in view)

   Which number will appear on the display after these keys are pressed?

   $2\ \boxed{\times}\ 3\ \boxed{=}\ 5\ \boxed{=}$

   a. 6   b. 5   c. 10   d. 30

Answer Key for Sample Tests

Test 1.

1. 605   **2.** 316   **3.** 750   **4.** 5133   **5.** 6601
6. $1.50   **7.** 357   **8.** 212   **9.** $2.53   **10.** $\boxed{\cdot}$
11. $\boxed{C}$ **12.** 20, 32, 47

Test 2. Questions 20–27 are to be answered without a calculator in view.

1. 908 591   **2.** 338 376   **3.** 10 178   **4.** 510 741
5. 180   **6.** 144 355 200   **7.** 3   **8.** 581   **9.** 3072
10. 5288   **11.** $10.40   **12.** $30.35   **13.** 11
14. 147   **15.** 10   **16.** 625   **17.** 16   **18.** 674
19. 68   **20.** c   **21.** c   **22.** b   **23.** b   **24.** c
25. c   **26.** b   **27.** b   **28.** d

Test 3. Questions 33–43 are to be answered without a calculator in view.

1. $81.40   **2.** 5 764 801   **3.** 85 dollars   **4.** 0.875
5. 37.5492   **6.** 496   **7.** 1826   **8.** 2142
9. 12.964912   **10.** 0.6666666 (or 2/3)   **11.** $40.14
12. 172 (round from 171.83333)   **13.** 1 202 798 040
14. 234   **15.** 32   **16.** $18.45   **17.** 15
18. 13 weeks   **19.** 26 and 25   **20.** 131 423 310 feet
21. July 30, $29.33; August 1, $38.22   **22.** 403
23. −86   **24.** −8.25   **25.** 81
26. 63 204 275 562 506   **27.** 72.8   **28.** $38.50
29. 40%   **30.** 2.5   **31.** 1 412 437.5
32. 511.5 inches   **33.** b   **34.** b   **35.** d   **36.** b
37. c   **38.** c   **39.** c   **40.** 28   **41.** c   **42.** d
43. a

Sample Test 1

Do each problem on a calculator. Clear your calculator for each problem. Write the answers in the boxes.

1. *Add* 358 and 247.

2. *Subtract* 386 from 702.

3. Find the *sum* of these four numbers:

87   139   462   62

4. *Multiply* 59 and 87.

5. *Add* 426 and 517. Then multiply the *sum* by 7.

6. Find the total value of these coins:

4 quarters

3 dimes

4 nickels

$ _____

Answer in dollars and cents

7. What is 714 minus 357?

8. Clear your calculator. Press these keys and write the calculator's answer.

92 ⊟ 16 ⊞ 47 ⊞ 89 ⊟

9. Suppose you had 5 dollars to spend and you spent $2.47. How much money do you have left?

Answer in dollars and cents

$ _____

10. There is a special key on the calculator that separates dollars from cents. Draw a circle around this key.

Ⓒ   ⊟   ⊡   ⊞

11. There is a special key that erases a problem from the calculator. Draw a circle around that key.

⊞   Ⓒ   ⊠   M+

12. Make your calculator count by pressing these keys: 8 ⊞ 3 ⊟ ⊟ ⊟ ⊟ . . . . Keep pressing ⊟ and look at the display. Circle all the numbers below that you see on the calculator display.

20   32   40   47   67

47

Sample Test 2

Do each problem on a calculator. Clear your calculator for each problem. Write the answers in the spaces provided.

1. What is 14 649 plus 893 942?

   _____

2. What is 358 000 minus 19 624?

   _____

3. Subtract 5893 from 16 071.

   _____

4. Multiply 847 and 603.

   _____

5. Divide 65 700 by 365.

   _____

6. Find the product of 5280 and 27 340.

   _____

7. Divide the sum of 64 and 287 by 117:

   $\dfrac{(64 + 287)}{117} =$ _____

8. Add the product of 24 and 18 to 149.

   _____

9. Multiply these three numbers. What is the product?

   $16 \times 8 \times 24 =$ _____

10. Find the sum of these four numbers:

    287   93

    621   4287

    _____

11. Suppose you buy 16 candy bars. The price of each candy bar is 65 cents. What is the total cost? (Write your answer in dollars and cents.)

    _____

12. Find the total for this grocery bill.

    April 27

    | Meat | $17.95 |
    | Vegetables | 8.07 |
    | Dairy | 3.16 |
    | Tax | 1.17 |
    | **Total** | _____ |

13. Joe works at the bakery. He packs cookies in boxes. If he wants to pack 6455 cookies in boxes of 36 cookies each, how many cookies will be left over?

    _____

14. Five thousand sheets of paper will be shared equally by 34 students. How many sheets of paper will each student get? (Answer with a whole number.)

    _____

15. How many times do you press $=$ to make the calculator start at 13 and count by 29s to reach 303?

    13 $+$ 29 $=$ $=$ . . .

    How many times?

    _____

16. Use 5 as a factor 4 times. Write the display.

    5 $\times$ $=$ $=$ $=$

    _____

17. Subtract 17 from 152 eight times. Write the display.

    152 $-$ 17 $=$ $=$ . . .

    _____

18. Enter 57 $+$ 82 $+$ 673 $-$ 98 $-$ 40 $=$.

    Write the final display.

    _____

19. Enter 92 $+$ 84 $+$ 71 $+$ 56 $+$ 37 $=$ $\div$ 5 $=$.

    Write the final display.

    _____

Put the calculator away and out of view. Circle the letter for the correct answer.

20. Which of these keys is used to divide one number by another number?

a. $\boxed{\sqrt{}}$ b. $\boxed{\text{M-}}$ c. $\boxed{\div}$ d. $\boxed{=}$ e. $\boxed{\text{CE}}$

21. Which of these key sequences will make an *error* sign appear on the display?

a. $0 \boxed{-} 8 \boxed{=}$

b. $0 \boxed{\div} 8 \boxed{=}$

c. $99\ 000\ 000 \boxed{+} 1\ 000\ 000 \boxed{=}$

d. $5\ 000 \boxed{\times} 6\ 000 \boxed{=}$

e. None of the above

22. Suppose you have just entered a wrong number on your calculator. You want to clear this wrong number and not the whole problem. Which key should be used?

a. $\boxed{\text{MC}}$ b. $\boxed{\text{CE}}$ c. $\boxed{=}$ d. $\boxed{\text{M+}}$ e. $\boxed{\text{ON}}$

23. Which of these sequences shows the keys to press in order to see the multiples of 3 (3, 6, 9, 12, 15, . . .)?

a. $4 \boxed{-} 1 \boxed{=} \boxed{=} \boxed{=} \ldots$

b. $3 \boxed{+} \boxed{=} \boxed{=} \boxed{=} \ldots$

c. $3 \boxed{\times} 3 \boxed{=} \boxed{=} \boxed{=} \ldots$

d. $3 \boxed{\times} 1 \boxed{=} \boxed{=} \boxed{=} \ldots$

e. None of the above

24. Suppose you press these keys:

$24 \boxed{-} 6 \boxed{=} \boxed{=} \ldots$

How many times would you press $\boxed{=}$ in order to see 0 on the display?

a. 6   b. 1   c. 4   d. 18

e. None of the above

25. Suppose you entered this problem in the calculator exactly as shown from left to right:

$$3 \boxed{+} 5 \boxed{\times} 4 \boxed{=}$$

What number would show on the display?

a. 20   b. 23   c. 32   d. 12   e. 60

26. What is the largest whole number that your calculator will display?

a. 999 million, 999 thousand, 999

b. 99 million, 999 thousand, 999

c. 9 billion

d. 999 999

e. None of the above

27. Suppose you pressed these keys:

$\boxed{\text{CM}} 6 \boxed{\times} 2 \boxed{\text{M+}} 4 \boxed{\times} 2 \boxed{\text{M+}}$

Which key can you press to see the total?

a. $\boxed{\text{M+}}$ b. $\boxed{\text{MR}}$ c. $\boxed{\text{M-}}$ d. $\boxed{\text{CM}}$ e. $\boxed{=}$

28. Suppose you pressed these keys:

$\boxed{\text{C}} 9 \boxed{+} 3 \boxed{+} 4 \boxed{\text{CE}} 8 \boxed{=}$

What answer would the calculator show?

a. 6   b. 9   c. 13   d. 20   e. 21

Sample Test 3

Use a calculator for questions 1–32. Write your answers on a separate sheet of paper.

1. Suppose you purchase 44 packages of hot dogs. Each package costs $1.85. What is the total cost?

2. Calculate 7^8.

3. How much is $\frac{5}{12}$ of 204 dollars?

4. Change this fraction to a decimal.

   $\frac{21}{24}$

5. Find the product of 0.87 and 43.16.

6. Divide 6.2 by 0.0125.

7. $362 + (93 \times 17) - 117$

   = _____

8. $(81 \times 16) + (91 \times 18)$

   $- 792 =$ _____

9. $\dfrac{13 \times (84 + 19) - 600}{57}$

   = _____

10. $\dfrac{252 \times 715}{546 \times 495} =$ _____

11. Suppose you purchased 6 pizzas at $4.95 each and 12 cans of soft drink at 87 cents each. How much was the total? (Use [M+] and [MR] keys.)

12. Find the average of these 6 bowling scores. Answer with a whole number.

167	213	192
112	204	143

13. What is the product of 97 432 and 12 345?

14. What is the square root of 54 756?

15. Suppose you are to pack 42 cookies in each box and you have 500 000 cookies to pack. How many cookies will be left over?

16. Jim purchased a new television. The selling price, including taxes and finance costs, was $256.40. Jim made a $35.00 down payment. He will pay the balance in 12 equal monthly payments. How much is each monthly payment?

17. Three students multiplied their ages and found that the product was 3150. The ages of two of the students were 14 and 15. What was the age of the third student?

18. You now have $29.37 in your coin jar. You started your savings with 12 cents and you have been saving $2.25 each week. For how many weeks have you been saving?

19. Which of the pairs of whole numbers that have a sum of 51 have the greatest product?

20. The diameter of the earth is 7927 miles. There are 5280 feet in one mile. Use this formula to calculate the distance around the earth. Answer in feet to the nearest whole number.

Distance around = 3.14 × diameter

21. Write the bank balance for July 30 and August 1.

		Beginning balance	$19.83
Deposit	**Date**	**Withdrawal**	**Balance**
$51.41	July 18		$71.24
	July 20	$27.72	43.52
	July 30	14.19	_____
8.89	August 1		_____

22–25. Solve for N.

22. $N = 15 \times 63 + {}^-542$ $N =$ _____

23. $-15N = 1290$ $N =$ _____

24. $16N = 112 - 244$ $N =$ _____

25. $\dfrac{N}{117} = \dfrac{36}{52}$ $N =$ _____

26. Although these numbers are too large to enter into the calculator, use your calculator to find the total.

$$
\begin{array}{r}
489\ 647\ 811\ 201 \\
4\ 640\ 973\ 577\ 496 \\
+\ 58\ 073\ 654\ 173\ 809 \\
\hline
\end{array}
$$

27. What is 16% of 455?

_____

28. $1.54 is 4% of how much money?

_____

29. $14.00 is what percent of $35.00?

_____

30. 0.45 is 18% of what number?

_____

31. $(1.5)^5 \times 186\ 000 =$ _____

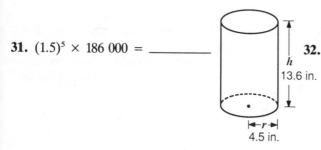
h
13.6 in.
|←—r—→|
4.5 in.

32. Find the surface area of a right circular cylinder with a radius of 4.5 inches and a height of 13.6 inches. Round your answer to the nearest tenth. Use 3.14 for pi.

Formula: Surface area $= 2\pi r^2 + 2\pi rh$

Put the calculator away and out of view.

33. These keys were pressed:

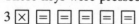
3 ✕ = = = = =

Which exponent was calculated?

a. 3^2 **c.** 3^5

b. 3^6 **d.** 5^3

34. Frank used his calculator to divide 1 by 99 million. What is true about his answer? 1 ÷ 99000000 =

a. 1 divided by 99 million is zero.

b. The calculator display shows zero, but the answer is not zero.

c. The calculator shows zero because Frank made a mistake.

d. All of the above are true.

35. What is the smallest nonzero positive number that can be shown on the calculator?

    a. 1.0   **c.** 0.000000001

    b. 0.1   **d.** 0.0000001

36. Look at the following calculator problem. What will the display show when the [MR] key is pressed?

    [MC] 16 [M+] 27 [M+] 18 [M−] [MR]

    a. 0   **b.** 25   **c.** 18   **d.** 43

37. There is a key on the calculator that will change the sign of the number from positive to negative or from negative to positive. Which of the following is that key?

    a. [−]   **b.** [=]   **c.** [+/−]   **d.** [M−]

38. Susan is multiplying 964 829 by 56 743 on her calculator. Her calculator display shows an error symbol. What has happened?

    a. She probably keyed in some incorrect numbers.

    b. She divided instead of multiplying.

    c. The product is too large for the 8-digit display.

    d. The error symbol came on accidentally.

39. Which problem below will cause an error symbol to appear on the calculator display?

    a. Dividing zero by a very large number

    b. Multiplying by zero

    c. Dividing a number by zero

    d. Pressing the times key when you want to do division

40. Suppose you pressed these keys. Write the calculator display.

    2 [+] 5 [×] 4 [=] _____

41–43. **Match the letter of the calculator solution with the correct problem.**

41. _____ $3.25 is 5% of what number?

42. _____ 5% of $3.25 is how much?

43. _____ What is 25% of 16?

    a. 16 [×] 25 [%]

    b. 3 [·] 25 [÷] 65 [%]

    d. 3 [·] 25 [÷] 5 [%]

    d. 3 [·] 25 [×] 5 [%]

Teacher Notes

COUNTING   Primary

Purpose: To reinforce counting by ones, twos, fives, and tens

Calculator Keys: $\boxed{+}$ [constant addend] $\boxed{=}$ $\boxed{C}$

Answer Key: Answers will vary. It takes about 20 to 30 seconds to make a calculator count by ones from 1 to 100. The children need some free time to explore counting on the calculator before they are ready to reflect on specific teacher questions about their counting.

COUNTING ON AND COUNTING BACK   Primary

Purpose: To count repetitions of the $\boxed{=}$ key

Calculator Keys: $\boxed{+}$ [constant addend] $\boxed{=}$ $\boxed{C}$

Answer Key:

1. 8   **2.** 10   **3.** 7   **4.** 5   **5.** 10   **6.** 7
7. 10   **8.** 6

MIXED ORDER   Upper Elementary

Purpose: To translate place values to numerals using the digit keys

Calculator Keys: $\boxed{+}$ $\boxed{=}$

Answer Key:

1. 532   **2.** 683   **3.** 502   **4.** 1252   **5.** 2060
6. 9023   **7.** 43   **8.** 126   **9.** 131   **10.** 208
11. 336   **12.** 189   **13.** 4433   **14.** 2376   **15.** 4325

NUMBER WORD NAMES   Upper Elementary

Purpose: To translate number word names to numerals

Calculator Keys: $\boxed{+}$ $\boxed{=}$

Answer Key: The exercises are self-checking.

CALCULATOR COUNTING PROBLEMS   Upper Elementary

Purpose: To foster readiness for division with a remainder

Calculator Keys: $\boxed{+}$ [constant addend] $\boxed{\times}$ $\boxed{=}$

Answer Key:

1. 13   **2.** 14   **3.** 17   **4.** 11   **5.** 18   **6.** 9

ADDITION AND SUBTRACTION   Primary

Purpose: To recognize addition and subtraction operations

Calculator Keys: $\boxed{+}$ $\boxed{-}$ $\boxed{=}$

Answer Key:

1. 133   **2.** 182   **3.** 51   **4.** 285   **5.** 84   **6.** 177   **7.** 78
8. 50

GUESS MY NUMBER (Subtraction)   Primary

Purpose: To reinforce mental computation and estimation skills in a game activity

Calculator Keys: [Constant addend] $\boxed{-}$ $\boxed{=}$

Answer Key: This game activity is self-checking.

ESTIMATING SUMS   Primary

Purpose: To encourage estimation

Calculator Keys: $\boxed{+}$ $\boxed{=}$

Answer Key: (exact, estimate)

1. 589, about 600   **2.** 470, about 500
3. 1067, less than 1200   **4.** 501, less than 500
5. 102, about 100   **6.** 156, less than 160
7. 211, about 210

ESTIMATE THE QUOTIENT   Upper Elementary

Purpose: To estimate the whole-number quotient

Calculator Keys: $\boxed{\times}$ $\boxed{=}$ $\boxed{-}$

Answer Key: Answers will vary. The rounded whole-number quotients are as follows:

1. 159   **2.** 145   **3.** 7   **4.** 1778   **5.** 385   **6.** 1996
7. 23   **8.** 4435

DIVISION ESTIMATION GAME   Middle School

Purpose: To estimate the whole-number quotient from a variety of division exercises under timed competition in a game activity

Calculator Keys: $\boxed{\div}$ $\boxed{=}$

Answer Key: (actual answer and an estimation range)

Game 1

1. 721
Est. 700–799

2. 121.95121
Est. 100–199

3. 128.09433
Est. 100–199

4. 3620.8333
Est. 3000–3999

5. 10.627118
Est. 10–19

6. 9000
Est. 9000–9999

7. 21.339035
Est. 20–29

8. 249.30019
Est. 200-299

9. 22.857142
Est. 20–29

10. 357
Est. 300–399

Game 2

1. 133.33333
Est. 100–199

2. 357.96226
Est. 300–399

3. 19.636363
Est. 10–19

4. 2022.2222
Est. 2000–2999

5. 9.3646833
Est. 1–9

6. 482.71606
Est. 400–499

7. 85.880701
Est. 80–89

8. 498.46021
Est. 400–499

9. 56.460369
Est. 50–59

10. 224.25
Est. 200–299

GUESS MY NUMBER (Division)   Upper Elementary

Purpose: To reinforce mental computation and estimation skills in a game activity

Calculator Keys: $\boxed{\div}$ $\boxed{=}$ [constant divisor]

Answer Key: Answers will vary

Note: The game is simpler to play when the divisors are limited to one-, two-, or three-digit whole numbers.

PATTERNS   **Middle School**

Purpose: To recognize a pattern and predict a calculation using a
   pattern

Calculator Keys: $\div$ $\times$ $=$ $+$ (The first factor entered is kept
   constant.)

Answer Key: (fourth answer only)

1. 57057   **2.** 68068   **3.** 1485   **4.** 121   **5.** 9090
6. 444   **7.** 9876   **8.** 2244422   **9.** 1665   **10.** 88888
11. (All 9 digits will not show.)   **12.** 4999995   **13.** 0.1212121
   1 2 3 4 5 4 3 2 1
14. 0.012012   **15.** 0.4545454

INVERSE   **Upper Elementary**

Purpose: To use the inverse operation in checking the answer to
   a computation

Calculator Keys: $+$ $-$ $\times$ $\div$ $=$

Answer Key:

1. 46547   **2.** 15842   **3.** 30295 (The original division answer
   is incorrect.)   **4.** 18695360 ÷ 592 = 31580 or 18695360
   ÷ 31580 = 592   **5.** 556308   **6.** 1869536 ÷ 592 = 3158
   or 1869536 ÷ 315800 = 5.92 (The original answer is
   incorrect.)   **7.** 17   **8.** 359191   **9.** 23   **10.** 50328531
11. 95   **12.** 144648

COUNTING BY TENTHS AND   **Upper Elementary**
   HUNDREDTHS

Purpose: To reinforce some decimal numeration concepts through
   counting by 0.1 and 0.01

Calculator Keys: $+$ $=$ [constant addend]

Answer Key:

1. decimals written from 0.1 to 1.5 by tenths   **2.** 1   **3.** 1.2
4. one and four tenths   **5.** 0.1   **6.** 0.2   **7.** 1   **8.** 0.4
9. 1.2   **10.** 0.5   **11.** 2

WRITING RATIOS AS DECIMALS   **Upper Elementary**

Purpose: To use reasoning and mental computation in learning
   the decimal equivalents for certain common fractions (fifths,
   fourths, eighths, thirds, tenths, sixths)

Calculator Keys: $\div$ $=$ [constant divisor]

Answer Key:

1. 0.4   **2.** 0.6   **3.** 0.8   **4.** 1   **5.** 1.2   **6.** 1.4   **7.** 0.5
8. 0.75   **9.** 1   **10.** 1.25   **11.** 1.5   **12.** 1.75   **13.** 0.25
14. 0.375   **15.** 0.5   **16.** 0.625   **17.** 0.75   **18.** 0.875
19. 1   **20.** 1.125   **21.** 1.25   **22.** E   **23.** A   **24.** I
25. C   **26.** G   **27.** D   **28.** J   **29.** B   **30.** F   **31.** H

DECIMAL ESTIMATION   **Upper Elementary**

Purpose: To estimate products and missing factors with decimals

Calculator Keys: $\times$ $=$

Answer Key: (The final attempt may be a decimal answer.)

1. 42   **2.** 65   **3.** 6   Answers for problems 4–11 will vary
   within the following ranges: **4..** 157–187   **5.** 607–681
6. 19–21   **7.** 886–906   **8.** 183–186   **9.** 5971–6082
10. 43–63   **11.** 715–785

REMAINDERS   **Upper Elementary**

Purpose: To calculate the whole-number remainder with a calculator and use the remainder in solving a problem

Calculator Keys: $\div$ $=$ $\times$ $-$

Answer Key:

1. 12   **2.** 10 (One bus will carry 24.)
3. 28 571 (15 cherries will not be boxed.)   **4.** 6   **5.** 838
6. 142 857 full weeks; 1 day left over

AVERAGE   **Upper Elementary**

Purpose: To compute the arithmetic mean of a set of numbers

Calculator Keys: $=$ $\div$

Answer Key:

1. 118   **2.** 66   **3.** 12.7 inches (rounded)
4. 154.0 cm (rounded)   **5.** $18.49 (rounded)   **6.** 33.3 years
(rounded)   **7.** 67, 64, 7   **8.** 57.6 or 58 years   **9.** 202

MEMORY   **Upper Elementary**

Purpose: To recognize and use the memory keys on the calculator

Calculator Keys: $\times$ $+$ $-$ $\div$ $=$ M+ M- MR MC

Answer Key:

1. 214   **2.** 99   **3.** 420   **4.** 14   **5.** 170   **6.** 0   **7.** 14 975
8. 1.305   **9.** 3.487   **10.** 15   **11.** 1087   **12.** 365

SHOPPING   **Upper Elementary**

Purpose: To use the memory keys in calculating the total cost of
   purchased items

Calculator Keys: $\times$ $=$ M+ M- MR MC

Answer Key:

1. $5.67   **2.** $13.34   **3.** $4.16 (rounded)   **4.** $14.40
5. $1.67   **6.** $0.29

ORDERING FAST FOOD   **Middle School**

Purpose: To use the memory keys to compute sums and differ-
   ences of products in a problem-solving situation

Calculator Keys: M+ M- MR MC $\times$ $=$ (Clear the memory
   before each exercise.)

Answer Key:

1. $13.52   **2.** $11.65   **3.** $17.09   **4.** $17.45   **5.** $17.50
6. The total is $77.21; the average is $2.57 a person per day.

CALCULATOR STORY PROBLEMS   **Upper Elementary**

Purpose: To work several different examples based on the same
   story problem

Calculator Keys: $\times$ $\div$ $-$ $+$ $=$

Answer Key:

1. $7.50   **2.** $3.75   **3.** $1.50   **4.** $16.50   **5.** $19.40
6. $42.00   **7.** $5.00   **8.** $10.40   **9.** $0.56   **10.** $4.80
11. $4.00   **12.** $5.40

STORY PROBLEMS (Fit the Facts)   **Primary**

Purpose: To practice solving story problems and develop awareness in using numbers to make sense of quantitative relationships

Calculator Keys: $\boxed{+}$ $\boxed{-}$ $\boxed{\times}$ $\boxed{\div}$ $\boxed{=}$

Answer Key:

1. 259
218 (or 41)
41 (or 218)

2. 68
39 (or 29)
29 (or 39)

3. 365
80¢
$292.

4. 400
16 (or 25)
25 (or 16)

5. 318 (or 213)
531
213 (or 318)

6. 2190 (or 1825)
4015
1825 (or 2190)

7. 3
458
350
1158

8. 2
38
15
91

GUESS AND TEST WITH A CALCULATOR   **Middle School**

Purpose: To use a calculator in solving story problems and use the guess-and-test strategy

Calculator Keys: $\boxed{=}$ $\boxed{-}$ $\boxed{\times}$ $\boxed{\div}$ ($\boxed{M+}$, $\boxed{MR}$, and $\boxed{MC}$ can be used.)

Answer Key:

1. eighteen–22¢ and twenty-seven–14¢
2. 33 adult, 22 child   **3.** 13, 16, 17   **4.** $2.75
5. 4 large, 2 small   **6.** 186 − 142 = 44 feet
7. $0.17 a bottle

PROPERTIES OF OPERATIONS   **Upper Elementary**

Purpose: To explore the order property (commutativity), grouping property (associativity), and distributive property

Calculator Keys: $\boxed{+}$ $\boxed{-}$ $\boxed{\times}$ $\boxed{\div}$ $\boxed{=}$

Answer Key:

1. 781   **2.** 16; yes (You have the difference with a positive or negative symbol.)   **3.** Order does matter.
4. Order does not matter.   **5.** 242
6. 200, 226; grouping matters   **7.** 7980
8. 2, 72; grouping matters   **9.** Answers will vary. Grouping matters in division and subtraction but not in multiplication or addition.
10. c   **11.** a   **12.** d

USE REASONING TO FIND THE ANSWER   **Upper Elementary**

Purpose: To reinforce the order, grouping, and distributive properties of the arithmetic operations

Calculator Keys: $\boxed{+}$ $\boxed{-}$ $\boxed{\times}$ $\boxed{\div}$ $\boxed{=}$

Answer Key:

Game 1	*Game 2*	*Game 3*
1. 224 352	**1.** 300 594	**1.** 9445
2. 0	**2.** 14 994	**2.** 8 235 756
3. 7 965 810	**3.** 285 600	**3.** 9445
4. 224 352	**4.** 301 436	**4.** 972
5. 7 965 810	**5.** 285 600	**5.** 8 235 756
6. 3 million	**6.** 14 994	**6.** 8 235 756

Game 4	*Game 5*	*Game 6*
1. 3 838 125	**1.** 1800	**1.** 6
2. 3 838 125	**2.** 809 999	**2.** 1 342 374
3. 75	**3.** 2	**3.** 0
4. 89	**4.** 809 999	**4.** 1 342 374
5. 3 838 125	**5.** 89 900	**5.** 4730
6. 0	**6.** 809 999	**6.** 283 800

MY DEAR AUNT SALLY   **Middle School**

Purpose: To learn the order of operations (multiplication and division before addition and subtraction)

Calculator Keys: $\boxed{+}$ $\boxed{-}$ $\boxed{\times}$ $\boxed{\div}$ $\boxed{=}$

Note: Most nonscientific pocket calculators process operations as they are entered. Some calculators are designed to use algebraic logic, i.e., they perform multiplication and division before addition and subtraction.

Answer Key:

1–10. Results are given.   **11.** 15 × 18 + 10 = 964
12. 521 − (16 × 16) = 265
13. (127 + 92) − (112 + 83) = 24
14. 1325 ÷ (184 − 131) = 25

LINEAR EQUATIONS   **Upper Elementary**

Purpose: To solve simple linear equations using inverse operations

Calculator Keys: $\boxed{+}$ $\boxed{-}$ $\boxed{\times}$ $\boxed{\div}$ $\boxed{=}$

Answer Key:

1. 33 **2.** 47 **3.** 22 **4.** 159 **5.** 88 **6.** 159 **7.** 6
8. 52 **9.** 144 **10.** 7333 **11.** 5 **12.** 852 **13.** 809
14. 1475 **15.** 16 **16.** 38 **17.** 25 **18.** 125 **19.** 1035
20. 3901

PROPORTIONS   **Middle School**

Purpose: To solve a proportion with a calculator

Calculator Keys: $\boxed{=}$ $\boxed{-}$ $\boxed{\times}$ $\boxed{\div}$ $\boxed{=}$

Answer Key:

1. 15 **2.** 160 **3.** 624 **4.** 48 **5.** 256 **6.** 51 **7.** 154
8. 169 **9.** 182 **10.** 950 **11.** 99 **12.** 87 **13.** 59

EXPONENTS   **Middle School**

Purpose: To calculate expressions containing exponential notation

Calculator Keys: $\boxed{\times}$ $\boxed{=}$

Answer Key:

Note: Most calculators will retain the first factor as a constant. Repeated pressing of the $\boxed{=}$ key produces increasing powers of a given number.

1. 343 **2.** 1024 **3.** 256 **4.** 1000 **5.** 1331 **6.** 7776
7. 1024 **8.** 59 049 **9.** 1 048 576 **10.** 125 000
11. 2304 **12.** 704 969 **13.** 3.375 **14.** 33.1776
15. 0.6561 **16.** 0.03125 **17.** 613.10657
18. Answers will vary. A decimal between 0 and 1 can be thought of as a fraction whose denominator is greater than

its numerator. As the exponent increases, the denominator increases at a faster rate than the numerator, and thus the quotient decreases.

19. 2, 4, 8, 6, 2, 4, 8, 6, . . .   **20.** 3, 9, 7, 1, 3, 9, 7, 1, . . .
21. 4, 6, 4, 6, . . .   **22.** 7, 9, 3, 1, 7, 9, 3, 1, . . .
23. E = 15   **24.** E = 8   **25.** E = 6

INTEGERS   **Middle School**

Purpose: To explore computation with integers.
Calculator Keys: $\boxed{+}$ $\boxed{-}$ $\boxed{\times}$ $\boxed{\div}$ $\boxed{=}$ $\boxed{\pm}$
Answer Key:

1. −96   **2.** 35   **3.** −18   **4.** 60   **5.** 0   **6.** −69
7. a,b 70   **8.** a,b −40   **9.** a,b −16   **10.** add   **11.** 56
12. −24   **13.** Yes   **14.** −252   **15.** 450

PERIMETER   **Upper Elementary**

Purpose: To solve perimeter problems
Calculator Keys: $\boxed{+}$ $\boxed{-}$ $\boxed{\times}$ $\boxed{\div}$ $\boxed{=}$
Answer Key:
1. 202 cm   **2.** 291 cm   **3.** 1109 cm   **4.** 8.6 km
5. 5056 cm (or 50.56 m)   **6.** 4.3 km   **7.** 405 cm
8. 249 cm   **9.** 114 cm   **10.** 1552 cm   **11.** 45 miles

AREA   **Upper Elementary**

Purpose: To use area formulas with a calculator
Calculator Keys: $\boxed{+}$ $\boxed{-}$ $\boxed{\times}$ $\boxed{\div}$ $\boxed{=}$
Answer Key:
1. 1260 cm²   **2.** 17 480 cm²   **3.** 11 977 cm²   **4.** 168.75 cm²
5. 8734.8 cm²   **6.** 7056 cm²   **7.** 4704 cm²   **8.** 36 208 cm²

CIRCLES, SPHERES, CONES, AND   **Middle School**
   CYLINDERS

Purpose: To compute areas and volumes involving pi (π)
Calculator Keys: $\boxed{+}$ $\boxed{-}$ $\boxed{\times}$ $\boxed{\div}$ $\boxed{=}$
Note: Use a constant factor to compute a power.
Answer Key:
1. 706.5 cm²   **2.** 70 650 cm³   **3.** 23 550 cm³   **4.** 399 cm³

CHANGE A RATIO TO A PERCENT   **Middle School**

Purpose: To use a calculator in expressing a ratio as a percent
Calculator Keys: $\boxed{\cdot}$ $\boxed{\div}$ $\boxed{\%}$
Answer Key:
1. 75   **2.** 15   **3.** 90   **4.** 60   **5.** 37.5 rounded to 38
6. 45 girls, 55 boys   **7.** 207.1 or 207   **8.** 171.4 or 171
9. 0.7   **10.** 0.2

SEVERAL PERCENTS OF THE SAME   **Middle School**
   BASE

Purpose: To gain intuitive awareness of 90, 50, 10, and 1 percent of the same base amount
Calculator Keys: $\boxed{\times}$ $\boxed{\%}$
Note: Enter the base first to use it as a constant factor.
Answer Key:

1. $150.00	**2.** $40.00	**3.** $325.00	**4.** $37.50
30.00	8.00	65.00	7.50
3.00	0.80	6.50	0.75
270.00	72.00	585.00	67.50
5. $2.50	**6.** $213.00	**7.** $ 9.30	**8.** $0.42
0.50	42.60	1.86	0.084
0.05	4.26	0.186	0.0084
4.50	383.40	16.74	0.756
5.00	426.00	18.60	0.84
7.50	639.00	37.20	1.68

ESTIMATING A PERCENTAGE USING   **Middle School**
   THE SAME RATE

Purpose: To get quick feedback in estimating percents of money amounts
Calculator Keys: $\boxed{\times}$ $\boxed{\%}$
Note: Enter the rate factor first to use it as a constant factor.
Answer Key:
1. $67.50   **2.** $35.68   **3.** $113.31   **4.** $432.00   **5.** $7.20
6. $30.60   **7.** $30.60   **8.** $126.63   **9.** $22.22
10. $729.00   **11.** $0.55   **12.** $5.63

TWELVE IS FIFTY PERCENT OF   **Middle School**
   WHAT?

Purpose: To solve percent problems with a calculator
Calculator Keys: $\boxed{\times}$ $\boxed{\div}$ $\boxed{\%}$
Answer Key:
1. 35 years   **2.** $30   **3.** 30   **4.** 1 253 709 votes   **5.** 5
6. 105   **7.** $416.00   **8.** $238.81   **9.** $41.71   **10.** $27.48
11. $231.96   **12.** $0.72

ADD-ONS AND MARKUPS   **Middle School**

Purpose: To solve add-on and discount percent problems with a calculator
Calculator Keys: $\boxed{+}$ $\boxed{-}$ $\boxed{\%}$
Note: (1) Some calculators require the use of $\boxed{=}$ to accomplish the actual addition or subtraction. (2) An alternate keystroke on some calculators is 46.2 $\boxed{\times}$ 5 $\boxed{\%}$ $\boxed{+}$ $\boxed{=}$.
Answer Key:
1. $117   **2.** $126   **3.** $96.14   **4.** $117   **5.** $126
6. $96.14   **7.** $10.06 (rounded)   **8.** 378   **9.** $58.05
10. $20.46 (rounded)